B

Systems & Control: Foundations & Applications
Volume 4

Series Editor
Christopher I. Byrnes, Washington University

Peter Falb

Methods of Algebraic Geometry in Control Theory: Part I

Scalar Linear Systems and Affine Algebraic Geometry

1990

Birkhäuser
Boston · Basel · Berlin

Peter Falb
Division of Applied Mathematics
Brown University
Providence, Rhode Island 02912
USA

Library of Congress Cataloging-in-Publication Data
Falb, Peter L.
    Methods of algebraic geometry in control theory / Peter Falb.
        p.   cm.—(Systems & control ; v. 4)
    Includes bibliographical references (p.
    Contents: v. 1. Scalar linear systems and affine algebraic
geometry.
    ISBN 0-8176-3454-1 (v. 1. : alk. paper)
    1. Control theory.   2. Geometry, Algebraic.   I. Title.
II. Series.
QA402.3.F34   1990
629.8'312—dc20   90-223

Printed on acid-free paper.

ISBN 0-8176-3454-1
ISBN 3-7643-3454-1

Camera-ready copy prepared by the author using TeX.
Printed and bound by Edwards Brothers, Inc., Ann Arbor, Michigan.
Printed in the U.S.A.

9 8 7 6 5 4 3 2 1

Preface

Control theory represents an attempt to codify, in mathematical terms, the principles and techniques used in the analysis and design of control systems. Algebraic geometry may, in an elementary way, be viewed as the study of the structure and properties of the solutions of systems of algebraic equations. The aim of these notes is to provide access to the methods of algebraic geometry for engineers and applied scientists through the motivated context of control theory.

I began the development of these notes over fifteen years ago with a series of lectures given to the Control Group at the Lund Institute of Technology in Sweden. Over the following years, I presented the material in courses at Brown several times and must express my appreciation for the feedback (sic!) received from the students. I have attempted throughout to strive for clarity, often making use of constructive methods and giving several proofs of a particular result. Since algebraic geometry draws on so many branches of mathematics and can be dauntingly abstract, it is not easy to convey its beauty and utility to those interested in applications. I hope at least to have stirred the reader to seek a deeper understanding of this beauty and utility in control theory.

The first volume deals with the simplest control systems (i.e. single input, single output linear time-invariant systems) and with the simplest algebraic geometry (i.e. affine algebraic geometry). This represents the beginning of both system theory and algebraic geometry. The classical frequency domain methods for control system design were primarily for scalar linear systems and did not always extend readily to multivariable systems. So, in a way, this first volume provides the algebraic geometry for understanding the mathematical structure (in part) of classical scalar control systems. It also provides the foundation for going further mathematically by developing the basic results of affine algebraic geometry.

While affine algebraic geometry is quite satisfactory and natural for scalar systems, the study of multi-input, multi-output linear time-invariant control systems requires the introduction of projective algebraic geometry. Since an important role for algebraic geometry to play in control theory is the description of the structure of multivariable linear systems, it is quite necessary to go beyond the basic results of the first volume. Consequently, the second volume of these notes is devoted to the study of multivariable linear systems and projective algebraic geometry. A section, ("Interlude"), indicating in a brief and heuristic fashion the flavor of Part II, is included in Part I.

While there are many friends, colleagues, teachers, and students, to whom considerable thanks are owed, I should like especially to express my appreciation to the

Laboratory for Information and Decision Systems at M.I.T. and its Director, Sanjoy K. Mitter, for the use of a quiet office (without which this book would still not be finished.) Thanks are also due to Ann Kostant for the excellent computer preparation of the manuscript.

Finally, I dedicate this work to my daughters, Hilary and Alison.

Table of Contents

0. Introduction

The overall goal of these notes is to provide an introduction to the ideas of algebraic geometry in the motivated context of system theory. While there are a number of excellent mathematical works on algebraic geometry (e.g. [D-1], [H-2], [K-4], [M-3], [S-2] etc.), these books do not deal with applications to system theory and are not, by virtue of their abstraction, always as accessible to engineers and applied scientists as the potential utility of the concepts warrants. We seek to provide a bridge for those interested in applications.

While we do not assume considerable mathematical knowledge beyond the basics of linear algebra, some simple notions from topology, and the elementary properties of groups, rings and fields, (see e.g. [B-1], [J-1], [Z-3]), we do suppose some knowledge of system theory and applications such as might be provided by a course on linear systems. Thus, the motivation for such ideas as, say, controllability will not be considered in detail here nor will we discuss questions of design. In addition, we deal only with the algebraic side of the subject and not with the analytic (or differential geometric) side. The comparable analytic development is examined, for example, in [H-4], [H-6], etc. Basic material on system theory can be found, for example, in [A-1], [B-4], [K-2], [R-1], and [W-1].

Part I deals with scalar (i.e. single input, single output) linear systems and affine algebraic geometry. This represents, in essence, the beginning of both system theory and algebraic geometry and is the subject of this volume.

We begin by introducing the four representations of a scalar linear system, namely: (i) a strictly proper rational meromorphic function (the transfer function); (ii) a pair of relatively prime polynomials (the differential operator); (iii) the Hankel matrix (the impulse response); and, (iv) a triple of matrices (the state space). A key complex of ideas revolves around the transition theorems from one representation to another. The first transition theorem (Hankel's theorem 1.4) relates rationality of a meromorphic function to the finiteness of the rank of a Hankel matrix.

In order to establish the second transition theorem, which is known as the Laurent Isomorphism Theorem (theorem 8.12 and corollary 9.15), we require some algebraic geometry. The basic notion of algebraic set as the zeros of a collection of polynomials is introduced (section 4.). Viewing the algebraic sets as the closed sets, we can define the Zariski topology on the affine N-space $\mathbb{A}^N$. This is the natural topology for algebraic geometry. The crucial connection between the geometric notion of an algebraic set and the algebraic notion of an ideal is studied (section 5.). In particular, we prove both the Hilbert Basis Theorem and the Hilbert Nullstellensatz. These theorems allow us to develop the key idea of affine algebraic geometry which is to associate with an algebraic set the ring of regular functions on that set. This association provides the connection between algebra and geometry and plays a vital role throughout the development. We can then prove the Laurent Isomorphism Theorem.

We have, up to that point, examined the relationships between the transfer function, differential operator, and Hankel representations of a scalar linear system. We proved that these representations were "algebraically" the same. We next consider the state space representation and its relation to the other representations. The realization maps are introduced (section 10), their relationships studied, and the set $S_{1,1}^n$ of *linear systems of degree* n is defined. The concepts of controllability and observability and minimality are defined. In addition, the group action and equivalence resulting from coordinate changes in the state space are examined. This leads to a theorem on the existence of minimal realizations (theorem 10.18) and to the State Space Isomorphism Theorem (theorem 11.19) which essentially states that minimal realizations are "equivalent" under coordinate change with a "unique equivalence". The group action leads naturally to a consideration of products of affine algebraic sets.

We then turn our attention to studying in a more general way such group actions as coordinate change and the corresponding equivalence relations. Calling the equivalence classes, *orbits*, we see that the main problem consists of two parts: first, to *find* the orbits; and second, to *parameterize* the orbits via an algebraic object. Functions which are constant on orbits are called *invariants* and we show that the characteristic realization map $\mathfrak{R}_\chi$ is an independent complete abstract invariant (proposition 13.17). This, however, is not enough to have an appropriate parameterization for state space equivalence. The better concept is that of a "geometric quotient" and we then state the Geometric Quotient Theorem (theorem 14.20). This may be viewed as the transition theorem between the state space representation and the other representations of a scalar linear system. The various proofs we give for the Geometric Quotient Theorem require us to treat a number of important algebraic ideas.

In particular, we start with an analysis of the notion of dimension. We give a topological definition. Then, we introduce the transcendence degree and use it to define an algebraic notion of dimension. An examination of the crucial concept of integral dependence follows and culminates in the "Going Down" Theorem (theorem 16.40). We also prove the Noether Normalization Lemma (lemma 16.43). We can then show the equivalence of the topological and algebraic definitions of dimension. Ultimately, this allows us to prove the first part of the Geometric Quotient Theorem, namely, that the orbits are closed.

We then turn our attention to the second property of a geometric quotient which we speak of as being "open on invariant sets". One proof of this property leads us to consider fibers of morphisms (section 18). We prove a theorem (18.5) which says that the dimension of the fibers is not "too small". We then define finite morphisms which are intimately related to integral dependence and prove the "Going Up" theorem (theorem 18.11). By using the process of factoring a morphism through a finite morphism (which is essentially a geometric form of the Noether Normalization Lemma), we prove that the dimension of the fibers is "just right almost everywhere". Next we establish a theorem of Chevalley (theorem 18.19) which states that morphisms carry constructible sets into constructible sets. Finally we prove the result required for the second property of a geometric quotient.

The third property of a geometric quotient involves showing that the ring of invariants is indeed the coordinate ring of the quotient. We first give an elementary proof based on a "canonical form". The second proof involves invariants under the symmetric group and utilizes results from Appendix C. The third proof is essentially a translation of the methods of classical invariant theory to the system theory context and generalizes readily to the multivariable case. With these proofs, we complete the development of the Geometric Quotient Theorem (theorem 14.20).

Next we focus on the final major result of scalar linear system theory, namely: the pole placement theorem. We describe the state feedback group and its action on the space of linear systems. Then we solve the pole placement problem (and variants). In particular we show that the "coefficient assignment" map is surjective if and only if the system is controllable.

We have, at this point, established the four major algebro-geometric results of scalar linear system theory, namely: (i) that the Laurent map is an isomorphism; (ii) that minimal state space realizations exist and that any two such minimal realizations are uniquely isomorphic; (iii) that a "geometric quotient" for equivalence under coordinate change exists; and, (iv) that the "poles" are assignable under state feedback if and only if the system is controllable. In order to develop a similar theory for multivariable systems, we require a more general approach. With this in mind, we conclude Part I with an examination of "abstract affine varieties" (section 22) and an "interlude" (section 23) which indicates some of the flavor of Part II.

Several additional comments are in order here. First, we do not strive for the greatest generality nor always for the simplest proof of a result. Frequently, we use constructive methods with a view towards application and also give several proofs of a particular theorem. We believe this is helpful in an introductory work. Second, the exercises are an integral part of the treatment and are used in the main body of the text. Finally, we have, of course, drawn on many sources and we acknowledge their considerable contribution (e.g. [A-2], [D-1], [F-2], [H-2], [H-8], [M-2], [S-2], [Z-3], etc.) even if explicit mention of them is not made at a particular point in the text.

Conventions

All rings are assumed commutative with an identity element 1. A ring homomorphism maps 1 into 1. Neither an integral domain or a field is the zero ring (i.e. $0 \neq 1$) and consequently, a prime (or maximal) ideal is necessarily a proper ideal (i.e. is not the ring itself). The notation $A \subset B$ means A is contained in B and A may equal B while the notation $A < B$ means A is contained in B but A is not equal to B (in other words, A is strictly contained in B). For sets, the notation $A - B$ means the complement of B in A.

1. Scalar Linear Systems Over the Complex Numbers

Let us begin by considering single input, single output linear systems over the complex numbers, $\mathbb{C}$. Such systems may be characterized by either a transfer function $f(z)$, or a pair of relatively prime polynomials $p(z)$ and $q(z)$, or a Hankel matrix $H = (h_{i+j-1})_{i,j=1}^{\infty}$ (the impulse response), or a triple (A, b, c) where A is a matrix and b, c are vectors. More precisely, we have the following:

Definition 1.1.   A *scalar linear system over* $\mathbb{C}$ is any one of the following: (a) a strictly proper rational meromorphic function $f(z)$ (i.e. $f(\infty) = 0$) with Laurent series $f(z) = \sum_{j=1}^{\infty} h_j z^{-j}$ about ∞; (b) a pair of relatively prime polynomials $p(z) = b_0 + b_1 z + \cdots + b_{m-1} z^{m-1}$, $q(z) = a_0 + a_1 z + \cdots + a_{m-1} z^{m-1} + z^m$; (c) a matrix $H = (h_{i+j-1})_{i,j=1}^{\infty}$ with finite rank; and, (d) a triple (A, b, c) with A an $n \times n$ matrix (i.e. $A \in M(n, n; \mathbb{C})$), b an $n \times 1$ column vector (i.e. $b \in \mathbb{C}^n$) and c a $1 \times n$ row vector (i.e. $c \in \mathbb{C}^{n'}$).

The first question, which arises quite naturally, is: what are the relationships between these various concepts?

Clearly, if $f(z)$ is a strictly proper rational meromorphic function, then $f(z) = p(z)/q(z)$ with $p(z)$ and $q(z)$ relatively prime polynomials of the form given in (b) and conversely.

If (A, b, c) is a linear system then $\det[zI - A] = q_A(z)$ is a monic polynomial of degree n and $(zI - A)^{-1} = \sum_{j=1}^{n} \phi_j(z) A^{j-1}/q_A(z)$ where $\phi_j(z)$ is a polynomial of degree $n - j$. It follows that $f(z) = c(zI - A)^{-1} b = \sum_{j=1}^{n} \phi_j(z) c A^{j-1} b / q_A(z)$ is a strictly proper rational meromorphic function since $\sum_{j=1}^{n} \phi_j(z) c A^{j-1} b$ is a polynomial of degree less than n. However, it is not necessarily true that the polynomials $\sum_{j=1}^{n} \phi_j(z) c \, A^{j-1} b$ and $q_A(z)$ are relatively prime (this involves the notion of minimality which shall be introduced in section 10.).

Example 1.2.   Let $A = I_2$ be the 2×2 identity matrix and let $b = \begin{bmatrix} 1 \\ 0 \end{bmatrix}$, $c = [\,1 \quad 0\,]$. Then $q_A(z) = (z-1)^2$ and $\sum_{j=1}^{2} \phi_j(z) c A^{j-1} b = (z - 1)$.

If (A, b, c) is a linear system, then the Hankel matrix $H =$

$(cA^{i+j-2}b)^{\infty}_{i,j=1}$ has finite rank by virtue of the Cayley-Hamilton theorem. On the other hand, if $H = (h_{i+j-1})^{\infty}_{i,j=1}$ has finite rank n, then, we have, from the form of the Hankel matrix H, a (unique) recurrence relation

$$a_0 h_j + a_1 h_{j+1} + \cdots + a_{n-1} h_{j+n-1} + h_{j+n} = 0 \qquad (1.3)$$

with $a_0, a_1, \ldots, a_{n-1}$ in $\mathbb{C}$, $j = 1, \ldots$. If we let

$$A = \begin{bmatrix} 0 & 1 & 0 & \cdots & 0 \\ 0 & 0 & 1 & \cdots & 0 \\ \vdots & \vdots & & & \vdots \\ \vdots & \vdots & & & 1 \\ -a_0 & -a_1 & \cdots & \cdots & -a_{n-1} \end{bmatrix},$$

$$b = H_n \begin{bmatrix} 1 \\ 0 \\ \vdots \\ 0 \end{bmatrix}, \quad c = [1 \quad 0 \quad \cdots \quad 0]$$

where $H_n = (h_{i+j-1})^{n}_{i,j=1}$, then (A, b, c) is a linear system with $cA^{j-1}b = h_j$ for $j = 1, \ldots$.

The relation between strictly proper rational meromorphic f and finite rank Hankel matrices is given by the following theorem.

Theorem 1.4 (Hankel [G-1]).   *Let $f(z) = \sum_{j=1}^{\infty} h_j z^{-j}$ be a strictly proper meromorphic function and let $H_f = (h_{i+j-1})^{\infty}_{i,j=1}$. Then f is rational if and only if rank H_f is finite.*

Proof 1: Suppose that $f(z)$ is rational. Then $f(z) = p(z)/q(z) = (b_0 + b_1 z + \cdots + b_{n-1} z^{n-1})/(a_0 + a_1 z + \cdots + a_{n-1} z^{n-1} + z^n)$ with $p(z)$, $q(z)$ relatively prime. It follows that $p(z) = (\sum_{j=1}^{\infty} h_j z^{-j})q(z)$. Comparing the coefficients of z^{-j}, we get the recurrence relation

$$a_0 h_j + a_1 h_{j+1} + \cdots + a_{n-1} h_{j+n-1} + h_{j+n} = 0 \qquad (1.5)$$

for $j = 1, \ldots$. Hence, all columns of H_f are in the span of the first n columns and rank $H_f \leq n$ is finite.

Conversely, if rank $H_f = n$, then there is a unique recurrence relation of the form (1.5). Setting $q(z) = a_0 + a_1 z + \cdots + a_{n-1} z^{n-1} + z^n$, we deduce that $(\sum_{j=1}^{n} h_j z^{-j}) q(z) = p(z)$ is a polynomial of degree less than n and, hence, that $f(z) = p(z)/q(z)$ is rational. Moreover, $p(z)$ and $q(z)$ are relatively prime for if there were a common factor, then a recurrence relation of length less than n would exist for the h_j contradicting rank $H_f = n$.

Proof 2: Suppose that $f(z)$ is rational with $f(z) = p(z)/q(z) = (b_0 + b_1 z + \cdots + b_{n-1} z^{n-1})/(a_0 + a_1 z + \cdots + a_{n-1} z^{n-1} + z^n)$. Let

$$A = \begin{bmatrix} 0 & 1 & 0 & \cdots & 0 \\ 0 & 0 & 1 & \cdots & 0 \\ \vdots & \vdots & \vdots & & \vdots \\ \vdots & \vdots & \vdots & & 1 \\ -a_0 & -a_1 & \cdots & \cdots & -a_{n-1} \end{bmatrix}$$

$$b = \begin{bmatrix} 0 \\ 0 \\ \vdots \\ 0 \\ 1 \end{bmatrix}, \; c = \begin{bmatrix} b_0 & b_1 & \cdots & b_{n-1} \end{bmatrix}$$

Then $c(zI - A)^{-1} b = f(z)$ for if we let $x = (zI - A)^{-1} b$, then $(zI - A)x = b$, $x_1 = 1/\det [zI - A]$, and $zx_j = x_{j+1}$ for $j = 1, \ldots, n - 1$. But $c(zI - A)^{-1} b = z^{-1} c(I - A/z)^{-1} b = \sum_{j=1}^{\infty} (cA^{j-1} b) z^{-j} = f(z) = \sum_{j=1}^{\infty} h_j z^{-j}$ and so, $h_j = cA^{j-1} b$. By the Cayley-Hamilton theorem, $a_0 I + a_1 A + \cdots + A^n = 0$ so that $c(a_0 A^j + a_1 A^{j+1} + \cdots + A^{j+n}) b = 0$, $j = 0, 1, \ldots$, or, in other words, $a_0 h_j + a_1 h_{j+1} + \cdots + h_{j+n} = 0$ and rank H_f is finite.

Conversely, if rank H_f is finite, then, as before, there is a unique recurrence relation of the form (1.5). If we let

$$A = \begin{bmatrix} 0 & 1 & 0 & \cdots & 0 \\ 0 & 0 & 1 & \cdots & 0 \\ \vdots & \vdots & \vdots & & \vdots \\ \vdots & \vdots & \vdots & & 1 \\ -a_0 & -a_1 & \cdots & \cdots & -a_{n-1} \end{bmatrix}$$

$$b = H_n \begin{bmatrix} 1 \\ 0 \\ \vdots \\ 0 \end{bmatrix} = \begin{bmatrix} h_1 \\ h_2 \\ \vdots \\ h_n \end{bmatrix}, \quad c = \begin{bmatrix} 1 & 0 & \cdots & 0 \end{bmatrix}$$

then $cA^{j-1}b = h_j$, $j = 1, \ldots$. But $c(zI - A)^{-1}b = z^{-1}(I - A/z)^{-1}b = \sum_{j=1}^{\infty}(cA^{j-1}b)z^{-j} = \sum_{j=1}^{\infty} h_j z^{-j} = f(z)$. Since $(zI - A)^{-1} = \sum_{j=1}^{n} \phi_j(z)A^{j-1}/\det[zI - A]$ where $\phi_j(z)$ is a polynomial of degree $n-j$ and $cA^{j-1}b = h_j$, we deduce that $f(z) = c(zI - A)^{-1}b = \sum_{j=1}^{n} \phi_j(z)h_j/\det[zI - A]$. But $\sum_{j=1}^{n} \phi_j(z)h_j$ has degree less than n and so, $f(z)$ is rational.

Now, by way of summary, we have shown that the concepts (a), (b) and (c) of definition 1.1 are readily related and that there are natural ways to go from a triple (A, b, c) to a transfer function $f(z)$ (i.e. $f(z) = c(zI - A)^{-1}b$) and to a Hankel matrix H (i.e. $H = (cA^{i+j-2}b)_{i,j=1}^{\infty}$). Filling in the gaps will require additional ideas which we develop in the sequel. Also, we observe that the notions (b), (c) and (d) of definition 1.1 make sense over any field k and the various proofs are algebraic in that these proofs do not depend on convergence. To give an appropriate meaning to (a) and the general development, we require some algebraic ideas which we develop in the sequel.

2. Scalar Linear Systems Over a Field k

Based on the intuition developed for the complex number case, we now turn our attention to the situation over a field k. Our first task will be to give an algebraic analog of the notion of a strictly proper meromorphic function. To do this, we introduce formal power series.

Let R be a commutative ring with (multiplicative) identity 1. If $a, b \in R$, then we say that a *divides* b (or is a *divisor* of b) if there is a c in R such that $b = ac$. An element a is *irreducible* (or prime) if a is not a unit (i.e. a divisor of 1) and any divisor of a is of the form ae where e is a unit. R is an *integral domain* if there are no proper divisors of zero.

Let R be an integral domain. Consider the set of ordered pairs (r, s) with $r, s \in R$ and $s \neq 0$. Call two such pairs (r, s) and (r_1, s_1) *equivalent* if there is a $t \neq 0$ in R such that $t(r_1 s - r s_1) = 0$ and write $(r, s) \sim (r_1, s_1)$. It is easy to see that $\sim$ is an equivalence relation. For example, if $(r, s) \sim (r_1, s_1)$ and $(r_1, s_1) \sim (r_2, s_2)$, then $t(r_1 s - r s_1) = 0$ and $t'(r_1 s_2 - r_2 s_1) = 0$ with $t, t' \neq 0$ imply $(tt' s_1)(r s_2 - r_2 s) = (tt' s_2)(r s_1 - r_1 s) + (tt' s)(r_1 s_2 - r_2 s_1) = 0$ so that $(r, s) \sim (r_2, s_2)$, We let $K(R)$ be the set of equivalence classes written r/s and we define addition and multiplication in $K(R)$ in the natural way (e.g. $r/s \cdot r_1/s_1 = r r_1/s s_1$). $K(R)$ is then a field called the *quotient field of* R. We note also that R can be identified with the subring of elements $r/1$ of $K(R)$. For example, $K(\mathbb{Z}) = Q$ (i.e. the quotient field of the integers is the rationals).

Let $R[x]$ be the polynomial ring in one variable over R (i.e. elements of $R[x]$ are polynomials $\sum_{j=1}^{n} a_j x^j$ with coefficients a_j in R). Recall that $p(x) \in R[x]$ is *homogeneous of degree* r if $p(x) = ax^r$ with $a \in R$.

Definition 2.1. A *formal power series* in one variable over R is an infinite sequence $u = (u_0, u_1, \ldots, u_i, \ldots)$ of homogeneous polynomials u_i in $R[x]$ with u_i of degree i or 0.

If $u = (u_0, u_1, \ldots)$, $v = (v_0, v_1, \ldots)$ are formal power series, then we set

$$u + v = (u_0 + v_0, u_1 + v_1, \ldots)$$
$$uv = (w_0, w_1, \ldots)$$

(2.2)

where $w_r = \sum_{i+j=r} u_i v_j$. With these definitions, the set of all formal power series over R becomes a commutative ring with identity $(1, 0, \ldots, 0, \ldots)$. The ring of formal power series over R is denoted by $R[[x]]$. If $u = (u_0, u_1, \ldots,)$ is an element of $R[[x]]$ with $u \neq 0$, then the first index r with $u_r \neq 0$ is called the *order of* u, $\nu(u)$, and u_r is called the *initial form* of u. We note that $\nu(u + v) \geq \min \{\nu(u), \nu(v)\}$ and $\nu(uv) \geq \nu(u) + \nu(v)$. We write $u = \sum_{j=0}^{\infty} u_j$ for the formal power series u, or, since $u_j = a_j x^j$, $u = \sum_{j=0}^{\infty} a_j x^j$. Note that $R[x]$ is naturally (identified with) a subring of $R[[X]]$.

Definition 2.3. Let z^{-1} be an indeterminate over R. An element $f(z^{-1})$ of $R[[z^{-1}]]$ of positive order is called a *strictly proper meromorphic function on* R. Such an $f(z^{-1})$ is *rational* if there are relatively prime polynomials $p(z)$ and $q(z)$ in $R[z]$ such that $f(z^{-1}) = p(z)/q(z)$.

If $f(z^{-1})$ is a strictly proper meromorphic function on R, then $f(z^{-1}) = \sum_{j=1}^{\infty} h_j z^{-j}$ with $h_j \in R$ and, by abuse of notation, we write $f(z) = \sum_{j=1}^{\infty} h_j z^{-j}$ and we call $\sum_{j=1}^{\infty} h_j z^{-j}$ the (formal) *Laurent series of* $f(z)$.

Now let k be a field. We then have the following:

Definition 2.4. A *scalar linear system over* k is any one of the following: (a) a strictly proper rational meromorphic function $f(z)$ with Laurent series $f(z) = \sum_{j=1}^{\infty} h_j z^{-j}$, $h_j \in k$; (b) a pair of relatively prime polynomials $p(z) = b_0 + b_1 z + \cdots + b_{m-1} z^{m-1}$, $q(z) = a_0 + a_1 z + \cdots + a_{m-1} z^{m-1} + z^m$ in $k[z]$; (c) a Hankel matrix $H = (h_{i+j-1})_{i,j=1}^{\infty}$ of finite rank with $h_j \in k$; and, (d) a triple (A, b, c) with A an $n \times n$ matrix (i.e. $A \in M(n, n; k)$), b an $n \times 1$ column vector (i.e. $b \in k^n$), and c a $1 \times n$ row vector (i.e. $c \in k^{n'}$).

We observe that all that was done over $\mathbb{C}$ remains valid over the field k and, in particular, we have:

Theorem 2.5 (Hankel). *Let* $f(z) = \sum_{j=1}^{\infty} h_j z^{-j}$ *and let* $H_f = (h_{i+j-1})_{i,j=1}^{\infty}$. *Then* f *is rational if and only if* H_f *has finite rank.*

We call the rank of H_f, the *degree of f*, and we note that if rank $H_f = n$, then the rank of $(H_f)_{n+j}$, $j = 0, 1, \ldots$, is also n. In general, if $H = (h_{i+j-1})_{i,j=1}^{\infty}$, then we say that H has *degree n* if H has rank n. We now make the following:

Definition 2.6. Let $\mathrm{Rat}(n, k) = \{f(z) : f(z)$ a strictly proper rational meromorphic function over k of degree $n\}$ and let $\mathrm{Hank}(n, k) = \{H : H$ a Hankel matrix over k of degree $n\}$.

In view of Theorem 2.5, there is a natural map $L : \mathrm{Rat}(n, k) \to \mathrm{Hank}(n, k)$ given by $L(f) = H_f$ which is surjective.

Our goal will be to give an appropriate algebraic structure to both $\mathrm{Rat}(n, k)$ and $\mathrm{Hank}(n, k)$ and to show that the mapping L is an "isomorphism" of the algebraic structure. We begin by observing that, although on the face of it both $\mathrm{Rat}(n, k)$ and $\mathrm{Hank}(n, k)$ involve infinite objects, there are actually only a finite number of parameters required because of the degree n condition. For example, $f(z)$ is an element of $\mathrm{Rat}(n, k)$ if and only if there are elements $b_0, \ldots, b_{n-1}, a_0, \ldots, a_{n-1}$ in k such that the polynomials $p(z) = b_0 + b_1 z + \cdots + b_{n-1} z^{n-1}$, $q(z) = a_0 + a_1 z + \cdots + a_{n-1} z^{n-1} + z^n$ are relatively prime. In other words, $\mathrm{Rat}(n, k)$ can be viewed as a subset of the affine space $\mathbb{A}_k^{2n}$ $(= k^{2n})$. More precisely, $\mathrm{Rat}(n, k) = \{(b_0, \ldots, b_{n-1}, a_0, \ldots, a_{n-1}) \in k^{2n} : f(z) = (b_0 + b_1 z + \cdots + b_{n-1} z^{n-1})/(a_0 + a_1 z + \cdots + a_{n-1} z^{n-1} + z^n)$ a strictly proper rational meromorphic function of degree $n\}$. We can also view $\mathrm{Hank}(n, k)$ as a subset of the affine space $\mathbb{A}_k^{2n}$. For, if H has rank n, then, in view of the recurrence relation (1.5), H is an element of $\mathrm{Hank}(n, k)$ if and only if there are elements $h_1, \ldots, h_n, a_0, \ldots, a_{n-1}$ in k such that $h_{j+n} = -\sum_{i=0}^{n-1} a_i h_{i+j}$, $j = 1, \ldots$, and no recurrence of order less than n exists. Alternatively, H is an element of $\mathrm{Hank}(n, k)$ if and only if the linear equation

$$\begin{bmatrix} h_1 & h_2 & \cdots & h_n \\ h_2 & h_3 & \cdots & h_{n+1} \\ \vdots & \vdots & & \vdots \\ h_n & h_{n+1} & \cdots & h_{2n-1} \end{bmatrix} \begin{bmatrix} x_0 \\ x_1 \\ \vdots \\ x_{n-1} \end{bmatrix} = \begin{bmatrix} h_{n+1} \\ h_{n+2} \\ \vdots \\ h_{2n} \end{bmatrix} \qquad (2.7)$$

has a *unique* solution $x_j = -a_j$, $j = 0, 1, \ldots, n-1$ with the a_j in k and such that (1.5) is satisfied. More precisely, $\mathrm{Hank}(n, k) = \{(h_1, \ldots, h_{2n-1}, h_{2n}) \in k^{2n} : H_n = (h_{i+j-1})_{i,j=1}^n$ has rank n i.e. is non-singular}. With $\text{Rat}(n,k)$ and $\text{Hank}(n,k)$ viewed in this way, the map $L : \text{Rat}(n,k) \to \text{Hank}(n,k)$ given by

$$L(f) = (h_1, \ldots, h_{2n-1}, h_{2n}) \tag{2.8}$$

is called the *Laurent map*. We shall eventually show that the Laurent map is an isomorphism of algebraic structures.

We observe that $\text{Hank}(n,k) = A_k^{2n} - V(\theta)$ where $V(\theta) = \{(h_1, \ldots, h_{2n}) \in k^{2n} : \theta(h_1, \ldots, h_{2n-1}) = \det H_n = 0\}$. In other words, $\text{Hank}(n,k)$ is the complement of the "zero set" of a polynomial. Such "zero sets" of polynomials are the basic building blocks from which algebraic geometry is built. We shall show in the next section that $\text{Rat}(n,k)$ can also be described as the complement of the "zero set" of a polynomial (called the resultant of p and q).

If $\text{Rat}(n,k)$ and $\text{Hank}(n,k)$ are viewed as subsets of A_k^{2n}, then we can ask for an explicit description of the Laurent map L. Suppose that $(b_0, \ldots, b_{n-1}, a_0, \ldots, a_{n-1})$ is an element of $\text{Rat}(n,k)$ (i.e. $f(z) = (b_0 + b_1 z + \cdots + b_{n-1} z^{n-1})/(a_0 + a_1 z + \cdots + a_{n-1} z^{n-1} + z^n)$ is rational of degree n). Then, in view of the second proof of Hankel's Theorem, we have $L(f) = (h_1(f), \ldots, h_{2n}(f))$ where

$$h_j(f) = h_j(b_0, \ldots, b_{n-1}, a_0, \ldots, a_{n-1}) = cA^{j-1}b \tag{2.9}$$

and

$$A = \begin{bmatrix} 0 & 1 & 0 & \cdots & 0 \\ 0 & 0 & 1 & \cdots & 0 \\ \vdots & \vdots & \vdots & & \vdots \\ \vdots & \vdots & \vdots & & 1 \\ -a_0 & -a_1 & \cdots & \cdots & -a_{n-1} \end{bmatrix}, b = \begin{bmatrix} 0 \\ 0 \\ \vdots \\ 1 \end{bmatrix}, c = [b_0, \ldots, b_{n-1}]$$

$$\tag{2.10}$$

and $j = 1, \ldots, 2n$.

Example 2.11.   Let $n = 2$. Then the explicit formula (2.9) becomes

$$h_1 = b_1$$
$$h_2 = b_0 - a_1 b_1$$
$$h_3 = -a_1 b_0 - a_0 b_1 + a_1^2 b_1 = -a_0 h_1 - a_1 h_2$$
$$h_4 = -a_0 b_0 + a_1^2 b_0 + 2a_0 a_1 b_1 - a_1^3 b_1 = -a_0 h_2 - a_1 h_3$$

so that the h_j are polynomials in (b_0, b_1, a_0, a_1). If $J(h_1, \ldots, h_4;$ $b_0, b_1, a_0, a_1)$ is the Jacobian matrix of the partial derivatives of the h_j with respect to the b_i and a_i, then $\det J(h_1, \ldots, h_4; b_0, b_1, a_0, a_1) = b_0^2 - a_1 b_1 b_0 + a_0 b_1^2 = -\det H$ is the same as the determinant of either of the matrices

$$\begin{bmatrix} c \\ cA \end{bmatrix} = \begin{bmatrix} b_0 & b_1 \\ -a_0 b_1 & b_0 - a_1 b_1 \end{bmatrix} = \mathcal{O}(A, c)$$

$$\begin{bmatrix} b_0 & b_1 & 0 \\ 0 & b_0 & b_1 \\ a_0 & a_1 & 1 \end{bmatrix} = \rho((b_0, b_1), (a_0, a_1))$$

The significance of this observation will become clear in the sequel.

We can also give an explicit description of the map F (actually L^{-1}) from $\mathrm{Hank}(n, k)$ into $\mathrm{Rat}(n, k)$. Suppose that $(h_1, \ldots, h_{2n}) \in \mathrm{Hank}(n, k)$ and let $H = (h_{i+j-1})_{i,j=1}^{n}$ (so that $\det H \neq 0$). Then (cf. (2.7)) the linear equation $Hx = h$ where $x' = (x_0, x_1, \ldots, x_{n-1})$, $h' = (h_{n+1}, \ldots, h_{2n})$ has a unique solution $x_j = -a_j$, $j = 0, 1, \ldots, n - 1$. In fact, if $(H^+)^j$ denotes the j-th row of the adjoint of H, then

$$a_j = a_j(h_1, \ldots, h_{2n}) = -(H^+)^{j+1} h / \det H \tag{2.12}$$

for $j = 0, 1, \ldots, n - 1$. Motivated by the second proof of Hankel's Theorem, we let

$$A = \begin{bmatrix} 0 & 1 & 0 & \cdots & 0 \\ 0 & 0 & 1 & \cdots & 0 \\ \vdots & \vdots & & & \vdots \\ \vdots & \vdots & & & 1 \\ -a_0 & -a_1 & \cdots & \cdots & -a_{n-1} \end{bmatrix}, b = H \begin{bmatrix} 1 \\ 0 \\ \vdots \\ 0 \end{bmatrix}, c = [1 \quad 0 \quad \cdots \quad 0]$$

so that $cA^{j-1} b = h_j$ for $j = 1, \ldots$. We observe that $\det(zI - A) = a_0 + a_1 z + \cdots + a_{n-1} z^{n-1} + z^n = q(z)$ and that $c(zI - A)^{-1} b = p(z)/q(z) = \sum_{j=1}^{\infty} (cA^{j-1} b) z^{-j} = \sum_{j=1}^{\infty} h_j z^{-j}$ where $p(z) = b_0 + b_1 z + \cdots + b_{n-1} z^{n-1}$ has coefficients given by the relations

$$b_r = \sum_{j=1}^{n-r} h_j a_{j+r} \tag{2.13}$$

for $r = 0, 1, \ldots, n - 1$, and where we have set $a_n = 1$ for convenience. Since H has rank n, $f(z) = p(z)/q(z)$ is an element of $\mathrm{Rat}(n, k)$ and so the mapping $F : \mathrm{Hank}(n, k) \to \mathrm{Rat}(n, k)$ is given explicitly by

$$F(h_1, \ldots, h_{2n}) =$$
$$(b_0(h_1, \ldots, h_{2n}), \ldots, b_{n-1}(h_1, \ldots, h_{2n}), \ldots, a_{n-1}(h_1, \ldots, h_{2n}))$$
$$\tag{2.14}$$

where the a_i and b_j are given by (2.12) and (2.13).

Example 2.15.   Let $n = 2$. Then the explicit formula (2.14) becomes

$$a_0 = - (h_3^2 - h_2 h_4)/(h_1 h_3 - h_2^2)$$
$$a_1 = - (h_1 h_4 - h_2 h_3)/(h_1 h_3 - h_2^2)$$
$$b_0 = h_1 a_1 + h_2 = (2 h_1 h_2 h_3 - h_1^2 h_4 - h_2^3)/(h_1 h_3 - h_2^2)$$
$$b_1 = h_1$$

so that the a_i and b_i are polynomials in the h_j divided by (at most) $\det H$. If $J(b_0, b_1, a_0, a_1; h_1, h_2, h_3, h_4)$ is the Jacobian matrix of the partial derivatives of the b_i and a_i with respect to the h_j, then the determinant $\det J(b_0, b_1, a_0, a_1; h_1, h_2, h_3, h_4) = -1/\det H$ is the same as the determinant of either of the matrices

$$\left(\begin{bmatrix} c \\ cA \end{bmatrix} \right)^{-1} = \left(\begin{bmatrix} b_0 & b_1 \\ -a_0 b_1 & b_0 - a_1 b_1 \end{bmatrix} \right)^{-1} = (\mathcal{O}(A, c))^{-1}$$

$$\left(\begin{bmatrix} b_0 & b_1 & 0 \\ 0 & b_0 & b_1 \\ a_0 & a_1 & 1 \end{bmatrix} \right)^{-1} = -\rho((b_0, b_1), (a_0, a_1))^{-1}$$

(with the b_i and a_i as functions of the h_j). Again, this observation will become clearer as we progress.

3. Factoring Polynomials

Now, in order to provide an appropriate algebraic structure for $\text{Rat}(n,k)$, we need to have an algebraic criterion for the requirement that two polynomials $p(z)$ and $q(z)$ be relatively prime i.e. have no common factor. In fact, we would like to develop such a criterion in terms of the coefficents $(b_0,\ldots,b_{n-1},a_0,\ldots,a_{n-1})$ of the polynomials. We shall do this by introducing the "resultant".

Definition 3.1.   An integral domain R is a *unique factorization domain* if (i) every non-unit in R is a finite product of primes, and, (ii) the factorization is unique to within order and unit factors. A polynomial $P(x)$ in $R[x]$ is *primitive* if its coefficients have no (non-unit) common divisor.

If $f(x) \in R[x]$ where R is a unique factorization domain, then $f(x) = c(f) \, f_1(x)$ where $c(f) \in R$ and $f_1(x)$ is primitive. $c(f)$ is called the *content of* f.

Example 3.2.   Let $R = Z$ be the set of integers. The polynomial $2x^2 + 3x + 5$ in $Z[x]$ is primitive and if $f(x) = 12x^2 + 18x + 30$, then $c(f) = 6$ (is the greatest common divisor of the coefficients of $f(x)$).

Lemma 3.3. (Gauss) *Let R be a unique factorization domain. If $f(x)$, $g(x) \in R[x]$, then $c(fg) = c(f)c(g)$ and hence, the product of primitive polynomials is primitive.*

Proof: Since $f = c(f)f_1$, $g = c(g)g_1$ and $fg = c(f)c(g)f_1 g_1$, it is enough to show that the product $f_1 g_1$ of the primitive polynomials f_1, g_1 is primitive. If not, then there is a prime p in R which divides the coefficients of $f_1 g_1$. Let $f_1 = \sum \alpha_i x^i$, $g_1 = \sum \beta_j x^j$ and suppose $\alpha_{i_0}, \beta_{j_0}$ are the first coefficients which are not divisible by p (note f_1, g_1 primitive). Then the coefficient of $x^{i_0 + j_0}$ in $f_1 g_1$ is of the form $\alpha_{i_0} \beta_{j_0}$ + terms divisible by p. Since this coefficient is divisible by p, we obtain the contradiction that p divides α_{i_0} or β_{j_0}.

Corollary 3.4. *If $g(x)$ divides $\alpha f(x)$ with $\alpha \in R$ and g primitive, then $g(x)$ divides $f(x)$.*

Proof: If $\alpha f(x) = g(x)h(x)$, then $\alpha c(f) = c(g)c(h) = \epsilon c(h)$ where ϵ is a unit. Hence, α divides $h(x)$ i.e. $\alpha h_2(x) = h(x)$ (where $h_2(x) = (c(f)\epsilon^{-1})h_1(x)$ and $h(x) = c(h)h_1(x)$). Since $R[x]$ is an integral domain, $f(x) = g(x)h_2(x)$.

Lemma 3.5. (Euclidean Algorithm). *Let R be an integral domain. If $f(x), g(x)$ are elements of $R[x]$ with degrees m, n respectively and $m \geq n$, then there are uniquely determined polynomials $d(x), r(x)$ such that*

$$\gamma^{m-n+1} f(x) = d(x)g(x) + r(x) \qquad (3.6)$$

where $r(x)$ has degree less then n and γ is the leading coefficient of $g(x)$.

Proof: Let α be the leading coefficient of $f(x)$. Then $\gamma f(x) - \alpha x^{m-n} g(x)$ has degree less than m and so, by induction, there are polynomials $d_1(x), r_1(x)$ such that $\gamma^{m-n}(\gamma f(x) - \alpha x^{m-n} g(x)) = d_1(x) g(x) + r_1(x)$ with degree $r_1(x) < n$. Let $d(x) = \alpha \gamma^{m-n} x^{m-n} + d_1(x)$, $r(x) = r_1(x)$ to obtain (3.6). As for uniqueness, if $\gamma^{m-n+1} f(x) = d'(x)g(x) + r'(x)$, then $(d(x) - d'(x))g(x) = (r'(x) - r(x))$. Since $r'(x) - r(x)$ has degree less than n, $d(x) = d'(x)$ and $r'(x) = r(x)$.

Theorem 3.7. *If R is a unique factorization domain, then $R[x]$ is a unique factorization domain.*

Corollary 3.8. $k[x_1, \ldots, x_n]$ *is a unique factorization domain.*

Proof (of theorem): We shall prove property (i) first using induction on the degree. If $f(x)$ has degree 0, then $f(x)$ factors into primes in R. If $f(x)$ has degree n (with $n > 0$), then $f(x) = c(f)f_1(x)$ with $f_1(x)$ primitive. If $f_1(x)$ is not irreducible, then $f_1(x) = g(x)h(x)$ and neither $g(x)$ nor $h(x)$ is constant. But then both $g(x)$ and $h(x)$ have degree less than n and factor by induction. As for property (ii), it is enough to show that if a prime $p(x)$ divides $f(x)g(x)$, then $p(x)$ is a divisor of either $f(x)$ or $g(x)$. If $p(x)$ has degree 0, then $p(x)$ is a divisor of either $c(f)$ or $c(g)$ and hence, of either $f(x)$ or $g(x)$. If $p(x)$ has positive degree, then

---

* If $m < n$, then $f(x) = 0 \cdot g(x) + f(x)$.

$p(x)$ is primitive. We suppose that $p(x)$ is not a divisor of $f(x)$ and that $f(x)$ has degree m. We let $M = R[x]p(x) + R[x]f(x)$. If $a(x)$ is a non-zero element of M of lowest degree n with leading coefficient α, then, by lemma 3.5, $\alpha^{m-n+1}f(x) = d(x)a(x) + r(x)$ with degree $r(x) < n$ or $r(x) = 0$. Since $d(x)a(x)$ and $\alpha^{m-n+1}f(x)$ are in M, $r(x) = 0$. Let $a(x) = c(a)a_1(x)$ with $a_1(x)$ primitive. In view of corollary 3.4, $a_1(x)$ is a divisor of $f(x)$ and similarly, $a_1(x)$ is a divisor of $p(x)$. Since $p(x)$ is prime and and does not divide $f(x)$, $a_1(x)$ is a unit in $R[x]$ and so, is an element of R. Hence $a(x) = c(a)a_1 \in R$ and $a \in M$ so that $ag(x) = b_1(x)p(x)g(x) + b_2(x)f(x)g(x)$. It follows that $p(x)$ divides $ag(x)$ and, by corollary 3.4, that $p(x)$ divides $g(x)$.

We can now prove a lemma which allows us to determine whether or not two polynomials are relatively prime.

Lemma 3.9. *Let R be a unique factorization domain and let $f(x) = \alpha_m x^m + \alpha_{m-1}x^{m-1} + \cdots + \alpha_0$, $g(x) = \beta_n x^n + \beta_{n-1}x^{n-1} + \cdots + \beta_0$ be elements of $R[x]$ with α_m or $\beta_n \neq 0$. Then f, g have a non-constant common factor if and only if there are $F(x), G(x)$ in $R[x]$ such that*

$$f(x)G(x) = g(x)F(x) \tag{3.10}$$

with degree $F < m$ and degree $G < n$.

Proof: Let $f(x) = f_1^{m_1} \cdots f_r^{m_r}$, $g(x) = g_1^{n_1} \cdots g_s^{n_s}$ be prime factorizations of f and g. If (say) f_1 is a common factor, then we simply let $F(x) = f(x)/f_1(x)$ and $G(x) = g(x)/f_1(x)$. On the other hand, if there is no common factor and (say) $\alpha_m \neq 0$, then $f(x)G(x) = g(x)F(x)$ implies $f_i^{m_i}$ divides F for $i = 1, \ldots, r$. Therefore,

$$\text{degree } F \geq \sum_{i=1}^{r} m_i \text{ degree } f_i = m.$$

Now, let $F(x), G(x)$ be given by

$$\begin{aligned}
F(x) &= A_{m-1}x^{m-1} + \cdots + A_0 \\
G(x) &= B_{n-1}x^{n-1} + \cdots + B_0
\end{aligned} \tag{3.11}$$

Then the polynomials $f(x)$ and $g(x)$ will have a non-constant common factor if and only if there are $A_0, \ldots, A_{m-1}, B_0, \ldots B_{n-1}$ not all 0 such that

$$(\alpha_m x^m + \cdots + \alpha_0)(B_{n-1} x^{n-1} + \cdots + B_0) = \\ (\beta_n x^n + \cdots + \beta_0)(A_{m-1} x^{m-1} + \cdots + A_0) \tag{3.12}$$

or, equating coefficients,

$$\alpha_m B_{n-1} = \beta_n A_{m-1}$$
$$\alpha_{m-1} B_{n-1} + \alpha_m B_{n-2} = \beta_{n-1} A_{m-1} + \beta_n A_{m-2}$$
$$\vdots \tag{3.13}$$
$$\alpha_0 B_0 = \beta_0 A_0$$

Given $f(x)$ and $g(x)$, (3.13) is a set of $m + n$ *linear* equations in the $m + n$ "unknowns" $A_0, \ldots, A_{m-1}, B_0, \ldots, B_{n-1}$. The coefficient matrix of these equations is

$$\begin{bmatrix} \beta_0 & 0 & \cdots & -\alpha_0 & 0 & \cdots & 0 \\ \beta_1 & \beta_0 & \cdots & -\alpha_1 & -\alpha_0 & & \vdots \\ \vdots & \vdots & & \vdots & & & \vdots \\ \beta_n & \vdots & & \vdots & & & \vdots \\ 0 & \beta_n & \cdots & -\alpha_m & & & \vdots \\ \vdots & & & \vdots & & & \vdots \\ 0 & \cdots & \beta_n & 0 & \cdots & \cdots & -\alpha_m \end{bmatrix}$$

and so, the system (3.13) will have a non-trivial solution if and only if the determinant of the following matrix is not zero:

$$\rho(f,g) = \begin{bmatrix} \alpha_0 & \alpha_1 & \cdots & \alpha_m & \cdots & \cdots & \cdots \\ 0 & \alpha_0 & \cdots & \cdots & \alpha_m & \cdots & \cdots \\ 0 & \cdots & \cdots & \cdots & \cdots & \alpha_m \\ \beta_0 & \beta_1 & \cdots & \cdots & \beta_n & \cdots & \cdots \\ 0 & \beta_0 & \cdots & \cdots & \cdots & \beta_n & \cdots \\ 0 & \cdots & \cdots & \cdots & \cdots & & \beta_n \end{bmatrix} \begin{matrix} \left. \begin{matrix} \\ \\ \\ \end{matrix} \right\} n \text{ rows} \\ \left. \begin{matrix} \\ \\ \\ \end{matrix} \right\} m \text{ rows} \end{matrix} \tag{3.14}$$

The determinant of 3.14 is called the *resultant of f and g* and is denoted by $\text{Res}(f,g)$.

Corollary 3.15. $f(x)$ and $g(x)$ have a non-constant common factor if and only if $Res(f,g) = 0$.

Now, returning to the system theory context, we can see immediately that if $f(z) = p(z)/q(z)$ then $f(z)$ will be a strictly proper rational meromorphic function if and only if $Res(p, q) \neq 0$. In other words, if $f(z) = p(z)/q(z)$ and H_f is the Hankel matrix generated by f, then $p(z)$ and $q(z)$ are relatively prime if, and only if, H_f has the right rank. This result may even be viewed as a system-theoretic treatment of the resultant. If $p(z) = b_0 + b_1 z + \cdots + b_{n-1} z^{n-1}$ and $q(z) = a_0 + a_1 z + \cdots + z^n$, then

$$
\text{Res}(p,q) = \det \begin{bmatrix} b_0 & b_1 & \cdots & b_{n-1} & \cdots & \cdots \\ 0 & b_0 & \cdots & \cdots & b_{n-1} & \cdots \\ \vdots & \vdots & & \vdots & \vdots & \\ a_0 & a_1 & \cdots & a_{n-1} & 1 & \cdots \\ 0 & a_0 & \cdots & \cdots & a_{n-1} & 1 \\ \vdots & & \ddots & \ddots & \ddots & \ddots \end{bmatrix} \begin{array}{l} \Big\} \ n \text{ rows} \\ \\ \Big\} \ n-1 \text{ rows} \end{array} \qquad (3.16)
$$

so that $\text{Res}(p,q) = \psi(b_0, \ldots, b_{n-1}, a_0, \ldots, a_{n-1})$ is in fact, a polynomial in the coefficients of the polynomials p and q. Thus, $\text{Rat}(n,k) = \mathbb{A}_k^{2n} - V(\psi)$ is the complement of the "zero set" of a polynomial. Such "zero sets" of polynomials are, as we shall soon see, the basic building blocks of affine algebraic geometry.

We have given a definition of a scalar linear system over k (see definition 2.4). Let us denote an "abstract" scalar linear system by the symbol Λ. The form $f(z)$ of Λ is called the *transfer function representation of* Λ; the form $(p(z), q(z))$ is called the *polynomial* (or *differential operator*) *representation of* Λ; the form $H = (h_{i+j-1})_{i,j=1}^{\infty}$ is called the *Hankel* (or *input-output* or *impulse response*) *representation of* Λ; and the form (A, b, c) is called a *state space representation* of Λ. We have shown that the transfer function and polynomial representations are indeed essentially the same and that these representations of degree n are the elements of $\text{Rat}(n, k)$. The Laurent map L, (2.8), gives a correspondence between the transfer function (or polynomial) representation and the Hankel representation. In fact, L gives a correspondence between systems of degree n. Investigation of this map will involve additional algebraic ideas. The relation between the state space representation and the other representations will be explored after our algebraic digression.

4. Affine Algebraic Geometry: Algebraic Sets

Let k be a field and let $\mathbb{A}_k^N = \{a = (a_1, \ldots, a_N) : a_i \in k\}$ be the *affine N-space over k*.

Definition 4.1. $V \subset \mathbb{A}_k^N$ is an *affine algebraic set* if $V = \{a = (a_1, \ldots a_N) : f_\alpha(a) = 0, f_\alpha(x) \in k[x_1, \ldots, x_N]\}$. In other words, V is an affine algebraic set if V is the set of common zeros of a family of polynomials.

Example 4.2.   Let $n = 2$ and $\psi(b_0, b_1, a_0, a_1) = b_0^2 - a_1 b_0 b_1 + a_0 b_1^2$. Then $V(\psi) = \{(b_0, b_1, a_0, a_1) \in \mathbb{A}_k^4 : \psi(b_0, b_1, a_0, a_1) = 0\}$ and so, for example, $(1, 1, 0, 1) \in V(\psi)$. In system theory terms, this means that since $p(z) = 1 + z$ and $q(z) = z + z^2 = z(z + 1)$ are not relatively prime, the degree of the transfer function $1 + z/z + z^2$ is one, not two.

Example 4.3.   Let $N = 2$ and $f(x_1, x_2) = x_1^3 - x_2^2$. Then $V(f) = \{(a_1, a_2) : a_1^3 = a_2^2\}$ is a semi-cubic parabola in $\mathbb{A}_k^2$. Note that the mapping $t \to (t^2, t^3)$ of $\mathbb{A}_k^1$ into $\mathbb{A}_k^2$ is a bijective map of $\mathbb{A}_k^1$ onto $V(f)$.

We shall in parts I and II assume that the field k is what is called "algebraically closed" and now work toward making that notion precise.

Let K be an extension field of k (i.e. $k \subset K$). An element ξ of K is *algebraic over k* if there is a non-zero polynomial $g(x)$ in $k[x]$ such that $g(\xi) = 0$ (i.e. ξ is a *root* of g). If every element of K is algebraic over k, then K is an *algebraic extension of k*. We observe that if $\xi \in K$ is algebraic over k, then there is a (unique) monic polynomial $m_\xi(x)$ of least degree such that $m_\xi(\xi) = 0$ and hence, $k[\xi]$ is a finite dimensional vector space over k. Conversely, if the dimension of $k[\xi]$ over k is finite, then ξ is algebraic over k as $1, \xi, \ldots, \xi^n, \ldots$ are linearly dependent over k. Moreover, if ξ is algebraic over k and $m_\xi(x) = x^n + a_{n-1}x^{n-1} + \cdots + a_0$ with $a_i \in k$ and (necessarily) $a_0 \neq 0$, then $\xi(-a_0^{-1}\xi^{n-1} - a_0^{-1}a_{n-1}\xi^{n-2} - \cdots - a_0^{-1}a_1) = 1$ and so $k[\xi]$ is a field $k(\xi)$. Suppose that $k \subset K \subset L$ and that K is a finite dimensional vector space over k with basis $u_1, \ldots, u_n$ and L is a finite vector space over K with basis $v_1, \ldots, v_m$. Then L is a finite dimensional vector space over k with basis $u_i v_j$, $i = 1, \ldots, n$, $j = 1, \ldots, m$. It follows that if K_1 is an algebraic extension of k and K_2 is an algebraic extension of K_1, then K_2 is an algebraic extension of k. This is called the *transitivity of algebraic extensions*.

Definition 4.4.   A field k is *algebraically closed* if every algebraic extension K of k coincides with k.

We shall *assume* that k is *algebraically closed*. For example, $k = \mathbb{C}$, the complex numbers, but $k \neq \mathbb{R}$, the real numbers.

Definition 4.5.   An *ideal* $\mathfrak{a}$ of a ring R is a subset of R which is an additive subgroup and satisfies $R\mathfrak{a} \subset \mathfrak{a}$ (i.e. $f, g \in \mathfrak{a}$ implies $f - g \in \mathfrak{a}$ and $f \in \mathfrak{a}$, $h \in R$ imply $hf \in \mathfrak{a}$).

We note that ideals correspond to the kernels of homomorphisms. If $\mathfrak{a}$ is an ideal, then $R/\mathfrak{a}$ denotes the residue class ring modulo $\mathfrak{a}$. Elements of $R/\mathfrak{a}$ are usually denoted $\bar{r}$ where $r \in R$.

Definition 4.6.   Let $\mathfrak{a}$ be an ideal in $k[x_1, \ldots, x_N]$ and let $V(\mathfrak{a}) = \{a = (a_1, \ldots, a_n) : f(a) = 0 \text{ for all } f \text{ in } \mathfrak{a}\}$. $V(\mathfrak{a})$ is called the *zero set* of the ideal $\mathfrak{a}$. If $W \subset \mathbb{A}_k^N$, then $I(W) = \{f \in k[x_1, \ldots, x_n] : f(W) = 0\}$ is called the *ideal of* W (clearly, $I(W)$ is an ideal).

Example 4.7.   Let $\mathfrak{a} = k[x_1, x_2] \cdot (x_1 x_2 - 1)$ (i.e. $\mathfrak{a}$ is the ideal consisting of all multiples of $x_1 x_2 - 1$). Then $V(\mathfrak{a}) = \{(a_1, a_2) : a_1 a_2 = 1\}$ is a "hyperbola".

Example 4.8.   If $W = (0, 0)$ in $\mathbb{A}_k^2$, then $I(W) = \{f \in k[x_1, x_2] : f \text{ has no constant term}\}$. For example, $x_1 x_2 + x_2^3$ is in $I(W)$ but $(x_1 + 3)(x_2 + 4)$ is not.

We now have:

Proposition 4.9. *Let* $\mathfrak{a}, \mathfrak{a}_i$ *be ideals in* $k[x_1, \ldots, x_N]$ *and let* W, W_i *be subsets of* $\mathbb{A}_k^N$. *Then*

(1) $\mathfrak{a} \subset \mathfrak{a}_1$ *implies* $V(\mathfrak{a}) \supset V(\mathfrak{a}_1)$

(2) $W \subset W_1$ *implies* $I(W) \supset I(W_1)$

(3) $V(\Sigma \mathfrak{a}_i) = \cap V(\mathfrak{a}_i)$

(4) $I(\cup W_i) = \cap I(W_i)$

(5) $V(\mathfrak{a} \cap \mathfrak{a}_1) = V(\mathfrak{a}\mathfrak{a}_1) = V(\mathfrak{a}) \cup V(\mathfrak{a}_1)$

(6) $V(I(W)) \supset W$ *and* $I(V(\mathfrak{a})) \supset \mathfrak{a}$

(7) $V(I(W)) = W$ *if and only if* W *an affine algebraic set*

(8) $I(V(\mathfrak{a})) = \mathfrak{a}$ *if and only if* $\mathfrak{a} = I(W)$ *for some* W.

Proof: Many are easy and we will only do (3), (5), and (7).

(3) $V(\Sigma \mathfrak{a}_i) \subset \cap V(\mathfrak{a}_i)$ since $\mathfrak{a}_i \subset \Sigma \mathfrak{a}_j$. But if $f \in \Sigma \mathfrak{a}_i$, then $f = \Sigma f_j$ (a finite sum) with $f_j \in \mathfrak{a}_j$ and so, $f = 0$ on $\cap V(\mathfrak{a}_i)$.

(5) Since $\mathfrak{a}\mathfrak{a}_1 \subset \mathfrak{a} \cap \mathfrak{a}_1$ and $\mathfrak{a} \cap \mathfrak{a}_1$ is a subset of of both $\mathfrak{a}$ and $\mathfrak{a}_1$, $V(\mathfrak{a}) \cup V(\mathfrak{a}_1) \subset V(\mathfrak{a} \cap \mathfrak{a}_1) \subset V(\mathfrak{a}\mathfrak{a}_1)$. If $c \notin V(\mathfrak{a}) \cup V(\mathfrak{a}_1)$, then there are $f \in \mathfrak{a}, g \in \mathfrak{a}_1$ such that $(fg)(c) = f(c)g(c) \neq 0$. But $fg \in \mathfrak{a}\mathfrak{a}_1$ and so, $c \notin V(\mathfrak{a}\mathfrak{a}_1)$. Thus, $V(\mathfrak{a}\mathfrak{a}_1) \subset V(\mathfrak{a}) \cup V(\mathfrak{a}_1)$.

(7) Obviously, $V(I(W)) = W$ implies that W is an algebraic set. If W is an algebraic set, then $W = V(\mathfrak{a})$ for some ideal $\mathfrak{a}$ and $I(W) = I(V(\mathfrak{a})) \supset \mathfrak{a}$ so that $V(I(W)) \subset V(\mathfrak{a}) = W \subset V(I(W))$.

Definition 4.10. $W \subset \mathbb{A}_k^N$ is *(Zariski-) closed* if W is an affine algebraic set. An open subset of an affine algebraic set is called a *quasi-affine algebraic set*.

We observe that $\mathbb{A}_k^N$ is a topological space under Zariski closure. Intersections of closed sets are closed by (3); finite unions of closed sets are closed by (5); and, $\phi = V(k[x_1, \ldots, x_n])$ and $\mathbb{A}_k^N = V((0))$. We note that the Zariski topology is not Hausdorff.

The sets $\mathrm{Rat}(n, k)$ and $\mathrm{Hank}(n, k)$ are both Zariski open sets in $\mathbb{A}_k^{2n}$. Moreover, the Zariski topology provides a notion of continuity for maps such as the Laurent map.

Example 4.11.   Let $L : \mathbb{A}_k^2 \to \mathbb{A}_k^2$ be given by $L(x, y) = (x, -xy)$. We assert that L is continuous (in the Zariski topology). For, if $W = \{(u, v) : f_\alpha(u, v) = 0\}$ is closed and if we let $V = \{(x, y) : g_\alpha(x, y) = 0\}$ where $g_\alpha(x, y) = f_\alpha(x, -xy)$, then V is closed and $L^{-1}(W) = V$. To see that $L^{-1}(W) = V$ we note that if $(x, y) \in V$, then $f_\alpha(L(x, y)) =$

$f_\alpha(x, -xy) = g_\alpha(x, y) = 0$ so that $(x, y) \in L^{-1}(W)$ and that if $(x, y) \in L^{-1}(W)$, then $0 = f_\alpha(L(x, y)) = f_\alpha(x, -xy) = g_\alpha(x, y)$ so that $(x, y) \in V$. For instance, if $W = \{(u, v) : f(u, v) = u + v = 0\}$, then $g(x, y) = x - xy$ and $V = \{(x, y) : x(1 - y) = 0\}$ (note that here L^{-1} of the line W consists of two lines). As another illustration, let $W = \{(u, v) : uv - 1 = 0\}$. Then $V = \{(x, y) : x^2 y + 1 = 0\}$. Now, let $n = 1$ so that $\mathrm{Rat}(1, k) = \{(b_0, a_0) : \mathrm{Res}(b_0, a_0 + z) = b_0 \neq 0\}$ and $\mathrm{Hank}(1, k) = \{(h_1, h_2) : h_1 \neq 0\}$. We observe that L *restricted to* $\mathrm{Rat}(1, k)$ is precisely the Laurent map and that $L(\mathrm{Rat}(1, k)) = \mathrm{Hank}(1, k)$ and that the Laurent map is injective. Thus, for $n = 1$, the Laurent map is a continuous bijective map.

Recalling that $\mathrm{Rat}(n, k) = \mathbb{A}_k^{2n} - V(\psi)$ and that $\mathrm{Hank}(n, k) = \mathbb{A}_k^{2n} - V(\theta)$ where $\psi = \mathrm{Res}(p, q)$ and $\theta = \det H$ are single polynomials, we have:

Definition 4.12.   Let $V \subset \mathbb{A}_k^N$ be an affine algebraic set and let $f \in k[x_1, \ldots, x_n]$ with $f \notin I(V)$. The set $V_f = \{v \in V : f(v) \neq 0\}$ is called a *principal affine open subset of V*.

We shall show that the principal affine open sets form a base for the Zariski topology and that algebraic sets can be determined by a finite number of equations in the next section.

5. Affine Algebraic Geometry: The Hilbert Theorems

Now, let us consider the question of whether or not algebraic sets can be determined by a finite number of equations.

Definition 5.1.   An ideal $\mathfrak{a}$ in a ring R has a *finite basis* if there are $f_1, \ldots, f_n$ in $\mathfrak{a}$ such that if $f \in \mathfrak{a}$, then $f = \sum_{i=1}^{n} r_i f_i$ with the r_i in R. We then call $f_1, \ldots, f_n$ a *basis* of $\mathfrak{a}$ and write $\mathfrak{a} = (f_1, \ldots, f_n)$. If $n = 1$ so that $\mathfrak{a} = (f) = Rf$, then $\mathfrak{a}$ is a *principal ideal*. If every ideal in the ring R has a finite basis, then R is a *Noetherian ring*.

Proposition 5.2.   *A ring R is Noetherian if and only if every ascending chain $\mathfrak{a}_1 \subset \mathfrak{a}_2 \subset \cdots$ of ideals terminates after finitely many terms i.e. $\mathfrak{a}_\nu = \mathfrak{a}_{\nu+1} = \cdots$ (ascending chain condition).*

Proof: If R is Noetherian and $\mathfrak{a}_1 \subset \mathfrak{a}_2 \subset \cdots$, then $\mathfrak{a} = \cup \mathfrak{a}_i$ is an ideal. But $\mathfrak{a} = (f_1, \ldots, f_n)$ implies all f_i are elements of some $\mathfrak{a}_\nu$ and hence, $\mathfrak{a}_\nu = \mathfrak{a}_{\nu+1} = \cdots = \mathfrak{a}$. Conversely, if there is an ideal $\mathfrak{a}$ without a finite basis, then there exist $f_i \in \mathfrak{a}$, $i = 1, 2 \ldots$ such that $(f_1) < (f_1, f_2) < \cdots$ is an infinite ascending chain of ideals.

We can now prove the Hilbert Basis Theorem which allows us to conclude that any affine algebraic set is defined by a finite number of polynomial equations.

Theorem 5.3. (Hilbert Basis) *If R is Noetherian, then $R[x]$ is also Noetherian.*

Corollary 5.4. $k[x_1, \ldots, x_N]$ *is Noetherian.*

Proof (of theorem): Let $\mathfrak{a}$ be an ideal in $R[x]$ and let $\ell_i(\mathfrak{a}) = \{c:$ there exists $f \in \mathfrak{a}$ of degree i with leading coefficient c or $c = 0\}$. $\ell_i(\mathfrak{a})$ is an ideal in R for: $c_1, c_2 \in \ell_i(\mathfrak{a})$ imply $f_1 - f_2 = (c_1 - c_2)x^i + \cdots$ is in $\mathfrak{a}$ and $rf_1 = (rc_1)x^i + \cdots$ is in $\mathfrak{a}$ for r in R. Let $\ell = \cup \ell_i(\mathfrak{a})$ so that ℓ is an ideal in R (as $\ell_i(\mathfrak{a}) \subset \ell_{i+1}(\mathfrak{a})$). Then ℓ has a finite basis $r_1, \ldots, r_n$

and we let $f_i(x), i = 1, \ldots, n$ be elements of $\mathfrak{a}$ with leading coefficients r_i. Let $d = \max (\text{degree } f_i)$. If $f \in \mathfrak{a}$ with $s = \text{degree } f \geq d$, then $f = cx^s + \cdots$ with $c \in \ell$ and $c = \sum_{i=1}^{n} \alpha_i r_i$. It follows that $f(x) - \sum \alpha_i f_i(x) x^{s-d_i} \in \mathfrak{a}$ (where $d_i = \text{degree } f_i$) and has degree less than s. Continuing, we obtain $\phi_1(x), \ldots, \phi_\nu(x)$ in $(f_1, \ldots, f_n)$ such that $f - \Sigma \phi_j$ is an element of $\mathfrak{a}$ of degree less than d. If $\ell_i(\mathfrak{a}) = (r_{i1}, \ldots, r_{i\nu(i)})$, $i = 0, 1, \ldots, d-1$ and $f_{ij}(x)$ are the corresponding elements of $\mathfrak{a}$, then clearly $\mathfrak{a} = (f_1, \ldots, f_n, f_{01}, \ldots, f_{0\nu(0)}, \ldots, f_{d-1\nu(d-1)})$ has a finite basis.

Proposition 5.5. *The principal affine open sets* $(\mathbb{A}_k^N)_f$, *where* $f \in k[x_1, \ldots, x_N]$, *form a base for the Zariski topology.*

Proof: Let U be Zariski open so that $U = \mathbb{A}_k^N - V$ with V an affine algebraic set. Let $I(V) = (f_1, \ldots, f_r)$. Then $V = \cap_{i=1}^{r} V((f_i))$ (by (3) of proposition 4.9). We claim that $U = \cup_{i=1}^{r} (\mathbb{A}_k^N)_{f_i}$. If $v \in \cup_{i=1}^{r} (\mathbb{A}_k^N)_{f_i}$, then $f_i(v) \neq 0$ for some i and $v \notin V$. On the other hand, if $v \in U$, then $v \notin V$ and $f_i(v) \neq 0$ for some i so that $v \in \cup_{i=1}^{r} (\mathbb{A}_k^N)_{f_i}$.

Example 5.6. Let $X = (x_{ij})_{i,j=1}^2$ be a 2×2 matrix with indeterminate entries; let $Y = \begin{bmatrix} y_1 \\ y_2 \end{bmatrix}$ be a 2×1 vector with indeterminate entries; and, let $Z = (z_1 z_2)$ be a 1×2 vector with indeterminate entries. If $\gamma(X, Y, Z) = \det[Y \; XY] = x_{21} y_1^2 + (x_{22} - x_{11}) y_1 y_2 - x_{12} y_2^2$ and $\omega(X, Y, Z) = \det [Z' \; X'Z'] = x_{12} z_1^2 + (x_{22} - x_{11}) z_1 z_2 - x_{21} z_2^2$, then $(\mathbb{A}_k^8)_\gamma$, $(\mathbb{A}_k^8)_\omega$ and $(\mathbb{A}_k^8)_{(\gamma\omega)} = (\mathbb{A}_k^8)_\gamma \cap (\mathbb{A}_k^8)_\omega$ are principal open sets (as γ and ω are elements of $k[X, Y, Z] = k[x_{11}, x_{12}, x_{21}, x_{22}, y_1, y_2, z_1, z_2]$).

We now consider the following questions: (1) if $\mathfrak{a} \subset k[x_1, \ldots, x_N]$ is a proper ideal (i.e. $\mathfrak{a} \neq k[x_1, \ldots, x_N]$), is $V(\mathfrak{a})$ non-empty?, and (2) how can the ideals $I(W)$ for W an algebraic set be characterized i.e. when is $\mathfrak{a} = I(V(\mathfrak{a}))$? These two seemingly unrelated questions are both answered by the Hilbert Nullstellensatz.

Definition 5.7. If $\mathfrak{a}$ is an ideal in a ring R, then the *radical of* $\mathfrak{a}$, $\sqrt{\mathfrak{a}}$, is the set $\{f : f^m \in \mathfrak{a} \text{ for some } m\}$.

Clearly, $\sqrt{\mathfrak{a}}$ is an ideal. Moreover, if $\mathfrak{a}$ is an ideal in $k[x_1, \ldots, x_N]$, then $\mathfrak{a} \subset \sqrt{\mathfrak{a}} \subset I(V(\mathfrak{a}))$ since $f(x_1, \ldots, x_N) \neq 0$ at $\xi = (\xi_1, \ldots, \xi_N) \in \mathbb{A}_k^N$ implies $f^m(\xi) \neq 0$ for all m.

Lemma 5.8. *Let $\mathfrak{a}$ be a proper ideal in $k[x_1, \ldots, x_N]$. Then $V(\mathfrak{a}) \neq \phi$ (note k is algebraically closed).*

Let us assume, for the moment, that this lemma is true. We then have:

Theorem 5.9. (Hilbert Nullstellensatz) *If $\mathfrak{a}$ is an ideal in the ring $k[x_1, \ldots, x_N]$, then $\sqrt{\mathfrak{a}} = I(V(\mathfrak{a}))$.*

Proof: We must show that given $f, f_1, \ldots, f_r$ in $k[x_1, \ldots, x_N]$ such that f vanishes at every common zero of the f_i, then there is an m with $f^m = \Sigma g_j f_j$, $g_j \in k[x_1, \ldots, x_N]$. Let x_{N+1} be a new indeterminate and consider $f_1, \ldots, f_r$, $1 - x_{N+1} f$ in $k[x_1, \ldots, x_{N+1}]$. Then $V((f_1, \ldots, f_r, 1 - x_{N+1} f)) = \phi$ (by lemma 5.8) and

$$1 = \sum_{i=1}^{r} h_i(x, x_{N+1}) f_i(x) + h_{r+1}(x, x_{N+1})(1 - x_{N+1} f(x)) \quad (5.10)$$

Setting $x_{N+1} = 1/f(x)$ and multiplying by a factor f^m (to clear denominators), we get the result.

Now, let us return to lemma 5.8. Before proving the lemma, we introduce the following:

Definition 5.11. An ideal $\mathfrak{m}$ in a ring R is *maximal* if $\mathfrak{m} \neq R$ and $\mathfrak{m} \subset \mathfrak{a} \subset R$ implies $\mathfrak{a} = \mathfrak{m}$ or $\mathfrak{a} = R$.

We observe that every proper ideal is contained in a maximal ideal by Zorn's Lemma. We also observe that an ideal $\mathfrak{m}$ is maximal if and only if $R/\mathfrak{m}$ is a field (since the ideals in $R/\mathfrak{a}$ are in one-one correspondence with the ideals of R which contain $\mathfrak{a}$). Thus, lemma 5.8 will follow from:

Lemma 5.12. *Let $\mathfrak{m}$ be a maximal ideal in $k[x_1, \ldots, x_N]$. Then $V(\mathfrak{m}) \neq \phi$.*

Proof: Let $K = k[x_1, \ldots, x_N]/\mathfrak{m}$ and let $\xi_i = \overline{x}_i$ (the residue class modulo $\mathfrak{m}$). Then K is a field, $(\xi_1, \ldots \xi_N) \in K$ and $f \in \mathfrak{m}$ implies $f(\xi_1, \ldots, \xi_N) = 0$. Thus, it will be enough to show that the ξ_i are algebraic over k (as k is algebraically closed). This will follow from:

Lemma 5.13. ([Z-1]) *If L is a field and $L[z_1, \ldots, z_N]$ is a field, then the z_i are algebraic over L.*

Proof: For $N = 1$, if z_1 is not algebraic over L, then $L[z_1]$ is a polynomial ring in one variable which is not a field. So we use induction on N. Since $L[z_1, \ldots, z_N]$ is a field, $L(z_1) = L_1 \subset L[z_1, \ldots, z_N]$ and $L_1[z_2, \ldots, z_N] \subset L[z_1, \ldots, z_N]$. But L_1 is a field and so, by induction, $z_2, \ldots, z_N$ are algebraic over L_1. Thus, by the transitivity of algebraic extensions, we need only show that z_1 is algebraic over L. If not, then z_i algebraic over $L_1 = L(z_1)$ implies there is a polynomial $f(z_1)$ such that $f(z_1)z_i$ satisfies a monic equation over $L[z_1]$ (by clearing denominators). Hence, if $g \in L[z_1, \ldots, z_N]$, then $f(z_1)^m g$ satisfies a monic equation over $L[z_1]$ for some m. But z_1 is not algebraic over L so that $L[z_1]$ is a unique factorization domain. Hence, if $h \in L_1 \subset L[z_1, \ldots, z_N]$, then h would be of the form $h'(z_1)/f(z_1)^m$ which is clearly false (by virtue of Proposition 5.14 which follows).

Proposition 5.14. *Let R be a unique factorization domain and let $z = x/y$ be an element of $K(R)$ (the quotient field of R) with x and y relatively prime. If $z^n + r_{n-1}z^{n-1} + \cdots + r_0 = 0$ with $r_i \in R$, then $z \in R$ i.e. y is a unit in R.*

Proof: If $(x/y)^n + r_{n-1}(x/y)^{n-1} + \cdots + r_0 = 0$, then $x^n = -y(r_{n-1}x^{n-1} + \cdots + r_0 y^{n-1})$ means y is a divisor of x^n. But then x and y would have a common prime factor if y were not a unit.

Corollary 5.15. *M is a maximal ideal in $k[x_1, \ldots, x_N]$ if and only if $M = (x_1 - \xi_1, \ldots, x_N - \xi_n)$ with $\xi_i \in k$ (i.e. $M = I(\xi)$, $\xi = (\xi_1, \ldots, \xi_N) \in \mathbb{A}_k^N$).*

Proof: If $M = (x_1 - \xi_1, \ldots, x_N - \xi_N)$, then $k[x_1, \ldots, x_N]/M = k$, a field, and M is maximal. On the other hand, if M is maximal, then $K = k[x_1, \ldots, x_N]/M = k(\xi_1, \ldots, \xi_N) = k$ (by lemma 5.13). Hence, $k + M = k[x_1, \ldots, x_N]$ and $x_i = \xi_i + m_i$ with $m_i \in M$. Therefore, $(x_1 - \xi_1, \ldots, x_N - \xi_N) \subset M$. By the first part, $(x_1 - \xi_1, \ldots, x_N - \xi_N)$ is maximal.

Corollary 5.16. *If $\mathfrak{a}$ is an ideal in $k[x_1, \ldots, x_N]$ and $R = k[x_1, \ldots, x_N]/\mathfrak{a}$, then every maximal ideal $\mathfrak{m}$ in R is of the form $\mathfrak{m} =$*

$(\overline{x}_1 - \xi_1, \ldots, \overline{x}_N - \xi_N)$ *with* $\xi_i \in k$ *and* $\overline{x}_i =$ a-*residue of* x_i.

Corollary 5.15 means that points of A_k^N and maximal ideals of $k[x_1, \ldots, x_N]$ are essentially the same object. In other words, $A_k^N = \{(\xi_1, \ldots, \xi_N) : \xi_i \in k\} = \{M = (x_1 - \xi_1, \ldots, x_N - \xi_N) : M$ a maximal ideal in $k[x_1, \ldots, x_N]\}$ (this set is called the *maximal spectrum* of $k[x_1, \ldots, x_N]$ and is denoted $\mathrm{Spm}(k[x_1, \ldots, x_N])$). Since any maximal ideal M is the kernel of a k-homomorphism α_M of $k[x_1, \ldots, x_N]$ onto k and conversely, we see that A_k^N and the set $\mathrm{Hom}_k(k[x_1, \ldots, x_N], k) = \{\alpha : \alpha$ is a k-homomorphism of $k[x_1, \ldots, x_N]$ into $k\}$ are also essentially the same object. This point of view, as we shall ultimately see, is the most fundamental.

6. Affine Algebraic Geometry: Irreducibility

We saw in the last section that points and maximal ideals are essentially the same object. Now a point is clearly the "smallest" non-empty algebraic set and so cannot be broken-up into smaller algebraic sets. This leads us to the following:

Definition 6.1.   A non-empty subset V of a topological space X is *irreducible* if V is not the union of two proper closed subsets of V (i.e. $V \neq V_1 \cup V_2$ with $V_1 < V$, $V_2 < V$ and V_1, V_2 closed in V).

Example 6.2.   Let $V = V(\mathfrak{a})$ where $\mathfrak{a} = (X_1^2 - X_2^2)$ in $k[X_1, X_2]$. Then $V = V_1 \cup V_2$ where $V_1 = V(X_1 + X_2)$ and $V_2 = V(X_1 - X_2)$ so that V is not irreducible. The algebraic sets V_1 and V_2 are irreducible.

Proposition 6.3. (i) *V is irreducible if and only if any two nonempty open sets in V have a nonempty intersection;* (ii) *a nonempty open subset U of an irreducible V is irreducible and dense; and,* (iii) *if V is irreducible, then the closure $\bar{V}$ of V is also irreducible.*

Proof:
(i) If V is irreducible and U_1, U_2 are nonempty open sets in V, then $U_1 \cap U_2 = \phi$ would imply $V = V_1 \cup V_2$ where $V_1 = V - U_1$, $V_2 = V - U_2$ are proper closed sets in V. Conversely, if $V = V_1 \cup V_2$ with V_1, V_2 proper closed subsets, then $U_1 = V - V_1$, $U_2 = V - V_2$ are nonempty open subsets of V with $U_1 \cap U_2 = \phi$.

(ii) Let U_1, U_2 be nonempty open subsets of U. Then $U_1 \cap U$, $U_2 \cap U$ are nonempty open subsets of V and so, $U_1 \cap U_2 = (U_1 \cap U) \cap (U_2 \cap U) \neq \phi$. If U were not dense, then $(V - \bar{U}) \cap U = \phi$ would contradict the irreducibility of V.

(iii) If $\bar{V} = V_1 \cup V_2$, then $V = (V_1 \cap V) \cup (V_2 \cap V)$. Since V is irreducible, we have (say) $V \subset V_1$. But V_1 is closed in $\bar{V}$ implies $V_1 = \bar{V}$ so that V_1 is not a proper subset.

Recalling that an ideal $\mathfrak{p}$ in a ring R is *prime* if $fg \in \mathfrak{p}$ implies that either $f \in \mathfrak{p}$ or $g \in \mathfrak{p}$ (or equivalently, if $R/\mathfrak{p}$ is an integral domain), we have:

Theorem 6.4. *An algebraic set* $V \subset \mathbb{A}_k^N$ *is irreducible if and only if* $I(V)$ *is a prime ideal in* $k[x_1, \ldots, x_N]$.

Proof: If $I(V) = \mathfrak{p}$ is a prime ideal and $V = V(I(V)) = V(\mathfrak{p}) = V_1 \cup V_2$, then $\mathfrak{p} = I(V_1) \cap I(V_2)$ implies $\mathfrak{p} = I(V_1)$ (say) and $V(\mathfrak{p}) = V_1 = V(I(V_1))$. On the other hand, if V is irreducible and $fg \in I(V)$, then $V \subset V((fg)) = V(f) \cup V(g)$ implies that $V = (V(f) \cap V) \cup (V(g) \cap V)$ and that either $V = V(f) \cap V$ or $V = V(g) \cap V$ so that $V \subset V(f)$ or $V \subset V(g)$. It follows that either $f \in I(V)$ or $g \in I(V)$.

Corollary 6.5. $\mathbb{A}_k^N$ *is irreducible* (*since* $I(\mathbb{A}_k^N) = (0)$ *is prime in* $k[x_1, \ldots, x_N]$).

Corollary 6.6. $\mathrm{Rat}(n, k)$ *and* $\mathrm{Hank}(n, k)$ *are irreducible* (*since* $\mathrm{Rat}(n, k)$ *and* $\mathrm{Hank}(n, k)$ *are nonempty open subsets of* $\mathbb{A}_k^{2n}$).

Corollary 6.7. *If* V *is irreducible, then every principal affine open subset of* V *is irreducible.*

Corollary 6.8. *If* $V \subset \mathbb{A}_k^N$ *is irreducible, then the residue class ring* $k[x_1, \ldots, x_N]/I(V)$ *is an integral domain.*

We shall, for the time being, call an irreducible affine algebraic set an *affine variety* and an open subset of an affine variety, a *quasi-affine variety*. Thus, $\mathrm{Rat}(n, k)$ and $\mathrm{Hank}(n, k)$ are quasi-affine varieties.

Example 6.9. Let $GL(2, k) = \{g : g \in M(2, 2; k), \det g \neq 0\}$. Since $M(2, 2; k)$ can be identified with $\mathbb{A}_k^4$, $GL(2, k)$ may be identified with the principal affine open subset defined by the nonvanishing of the determinant and so, $GL(2, k)$ is a quasi-affine variety. More explicitly, if

$$\xi = \begin{bmatrix} \xi_{11} & \xi_{12} \\ \xi_{21} & \xi_{22} \end{bmatrix} \in M(2, 2; k)$$

then $(\xi_{11}, \xi_{12}, \xi_{21}, \xi_{22}) \in \mathbb{A}_k^4$ (and conversely). If we let $X = (X_{11}, X_{12}, X_{21}, X_{22})$ with the X_{ij} as indeterminates, then $GL(2, k) = (\mathbb{A}_k^4)_\Delta$ where $\Delta(X) = \Delta(X_{11}, X_{12}, X_{21}, X_{22}) = X_{11}X_{22} - X_{12}X_{21}$.

We now have:

Proposition 6.10. *Every affine algebraic set V can be represented as a finite union, $V = \cup V_i$, of affine varieties V_i. The representation is unique (to within order) if no V_i is superfluous.*

Proof: If false, then there is a V which is not decomposable. Such a V is reducible and so $V = V_1 \cup V_1'$ with $V_1 < V$, $V_1' < V$ and (say) V_1 not decomposable. Continuing we obtain a chain $V > V_1 > \cdots$ and a chain $I(V) < I(V_1) < \cdots$. But $k[x_1, \ldots, x_N]$ is Noetherian (Corollary 5.4) and we have a contradiction. Now suppose that

$$V = \cup V_i = \cup V_j'$$

are two decompositions in which no V_i or V_j' is superfluous. Then $V_i = V \cap V_i = \cup V_j' \cap V_i$ implies $V_i \subset V_j'$ for some j as V_i is irreducible. Similarly, $V_j' \subset V_s$ for some s and so, $V_i \subset V_j' \subset V_s$. But then $V_i = V_s$ (since no V_i is superfluous) and $V_i = V_j'$. In other words, the set $\{V_i\}$ and the set $\{V_j'\}$ are the same.

Corollary 6.11. *If V is an affine algebraic set, then $I(V) = \mathfrak{p}_1 \cap \cdots \cap \mathfrak{p}_r$ where the $\mathfrak{p}_i$ are prime ideals in $k[x_1, \ldots, x_N]$ and no $\mathfrak{p}_i$ is superfluous.*

The sets V_i of Proposition 6.10 are called the *irreducible components* (or simply *components*) *of* V. They play a role in dealing with the notion of dimension.

7. Affine Algebraic Geometry: Regular Functions and Morphisms I

Now the key idea of affine algebraic geometry is to associate with algebraic sets the "regular" functions on these sets. Let us consider A_k^N and let $x_1, \ldots, x_N$ be the coordinate functions i.e. $x_i : A_k^N \to A_k^1$ is the mapping given by $x_i(\xi) = \xi_i$ where $\xi = (\xi_1, \xi_2, \ldots, \xi_N) \in A_k^N$.

Definition 7.1. A ring R is a *k-algebra* if k is a subring of R. A k-algebra R is *finitely generated over* k is there are elements $r_1, \ldots, r_n$ in R such that $R = k[r_1, \ldots, r_n]$. A finitely generated k-algebra without nilpotent elements is called an *affine k-algebra.*

Now, if $x_1, \ldots, x_N$ are the coordinate functions on A_k^N, then $k[x_1, \ldots, x_N]$ is a finitely generated k-algebra isomorphic to the polynomial ring in N-variables. Let V be an affine algebraic set and let

$$k[V] = k[x_1, \ldots, x_N]/I(V) \tag{7.2}$$

i.e. we define an equivalence relation by setting f equivalent to g if $f - g \in I(V)$ and $k[V]$ is the set of equivalence classes. We note that $k[V] = k[u_1, \ldots, u_N]$ where $u_i = \overline{x}_i$ is the equivalence class of x_i so that $k[V]$ is a finitely generated k-algebra. In fact, $k[V]$ is an affine k-algebra since $I(V) = \sqrt{I(V)}$ (by Theorem 5.9) and therefore, $\overline{f}^n = 0$ implies $f^n \in I(V)$ imples $f \in \sqrt{I(V)} = I(V)$ so that $\overline{f} = 0$. The elements of $k[V]$ are polynomial functions on V.

Definition 7.3. $k[V]$ is called the *affine coordinate ring of* V. The elements of $k[V]$ are called *regular functions* (on V).

Example 7.4. Let $V \subset A_k^2$ be the affine algebraic set given by $V = \{(\xi_1, \xi_2) : \xi_2 = \xi_1^2\}$. Let $F(X, Y) = \Sigma a_{ij} X^i Y^j$ be a polynomial in $k[X, Y]$ and let f be the restriction of F to V so that $f(\xi_1, \xi_1^2) = \Sigma a_{ij} \xi_1^i \xi_1^{2j}$. Then $f \in k[V]$. For instance, if $F(X, Y) = Y - X$, then $f : V \to A_k^1$ is given by $f(\xi_1, \xi_1^2) = \xi_1^2 - \xi_1$. If $G(X, Y) = X^2 - X$, then

G defines the same element of $k[V]$ as $F(X,Y)$. Note that $k[V] = k[X,Y]/(Y - X^2) = k[\overline{X},\overline{Y}] = k[\overline{X}]$ since $\overline{Y} = \overline{X}^2$ and so $k[V] \cong k[\mathbb{A}_k^1]$.

One of the reasons that the ring $k[V]$ is important is that V can be recovered from it. More precisely, if we denote the set $\{\alpha : \alpha$ is a k-homomorphism of $k[V]$ into $k\}$ by $\operatorname{Hom}_k(k[V],k)$, then we have :

Theorem 7.5. $V \cong \operatorname{Hom}_k(k[V],k)$.

Proof: If $v \in V$, then let $\alpha_v : k[V] \to k$ be given by $\alpha_v(f) = f(v)$. Clearly, α_v is an element of $\operatorname{Hom}_k(k[V],k)$ (e.g. $\alpha_v(fg) = (fg)(v) = f(v)g(v) = \alpha_v(f)\alpha_v(g)$). So we define a map $\psi : V \to \operatorname{Hom}_k(k[V],k)$ by setting $\psi(v) = \alpha_v$. Let $\mathfrak{m}_v$ be the kernel of α_v. Then $\mathfrak{m}_v$ is a maximal ideal in $k[V]$ (since α_v is a k-homomorphism) and so, by corollary 5.16, has the form $M_\xi/I(V)$ where $M_\xi = (x_1 - \xi_1, \ldots, x_N - \xi_N)$ is a maximal ideal of $k[x_1, \ldots, x_N]$. It follows that $v = (\xi_1, \ldots, \xi_N)$ and that ψ is injective. On the other hand, if $\alpha \in \operatorname{Hom}_k(k[V],k)$ and $\mathfrak{m}_\alpha$ is the kernel of α, then $k[V]/\mathfrak{m}_\alpha = k$ is a field which means that $\mathfrak{m}_\alpha$ is a maximal ideal in $k[V]$. Letting $M_\alpha = \{f : f \in k[x_1, \ldots, x_N], \overline{f} \in \mathfrak{m}_\alpha\}$, we have $M_\alpha \supset I(V)$ and M_α is a maximal ideal in $k[x_1, \ldots, x_N]$ (since $k \subset k[x_1, \ldots, x_N]/M_\alpha \cong (k[x_1, \ldots, x_N]/I(V))/(M_\alpha/I(V)) = k[V]/\mathfrak{m}_\alpha = k$). But then $M_\alpha = (x_1 - \xi_1, \ldots, x_N - \xi_N)$ and $V(M_\alpha) = (\xi_1, \ldots, \xi_N) = v \in V$ so that $\alpha = \alpha_v$. In other words, ψ is surjective.

Corollary 7.6. $V \simeq \operatorname{Spm}(k[V]) = \{\mathfrak{m} : \mathfrak{m}$ *a maximal ideal in* $k[V]\}$.

We also have :

Proposition 7.7.   *If* $R = k[f_1, \ldots, f_N]$ *is an affine k-algebra, then* $R = k[W]$ *for some affine algebraic set* W.

Proof: Let $\psi : k[X_1, \ldots, X_N] \to R$ be the k-homomorphism given by $\psi(X_i) = f_i$ and let $I = \ker \psi$. Then $I = \sqrt{I}$ since $f^n \in I$ implies $\overline{f}^n = 0$ (in $k[X_1, \ldots, X_N]/I$) so that $\overline{f} = 0$ (as $R \cong k[X_1, \ldots, X_N]/I$ has no nilpotent elements) and $f \in I$. It follows that $R = k[W]$ where $W = V(I)$.

Another of the reasons $k[V]$ is important is that it allows us to define the notion of morphism for affine algebraic sets. First, let $f_1, \ldots, f_m$ be elements of $k[x_1, \ldots, x_N]$. Then the f_i define a polynomial mapping ϕ of A_k^N into A_k^m as follows : if $\xi = (\xi_1, \ldots \xi_N) \in A_k^N$, then $\phi(\xi) = (f_1(\xi), \ldots, f_m(\xi))$.

Definition 7.8.   If $V \subset A_k^N$ and $W \subset A_k^m$ are affine algebraic sets, then a mapping $\phi : V \to W$ is a *k-morphism* if, for every g in $k[W]$, the composite function $g \circ \phi$ is in $k[V]$.

Definition 7.9.   If $V \subset A_k^N$ and $W \subset A_k^m$ are affine algebraic sets, then a mapping $\phi : V \to W$ is *regular* if ϕ is the restriction to V of a polynomial mapping.

We shall show that the notions in these two definitions are the same. The key to doing so is the fact that the regular functions on an affine algebraic set are the restrictions of polynomial functions.

So, now let $V \subset A_k^N$, $W \subset A_k^m$ be affine algebraic sets and let g be a polynomial function on W (i.e. $g \in k[W]$). If $\phi : V \to W$ is a regular map, then $g \circ \phi$ is clearly a polynomial function on V. Thus, every regular map is a morphism and we have a mapping $\phi^* : k[W] \to k[V]$ given by

$$\phi^*(g) = g \circ \phi \qquad (7.10)$$

ϕ^* is called the "lifting of ϕ".

Proposition 7.11.   *If φ is regular, then φ^* is a k-homomorphism.*

Proof: Simply note that $\varphi^*(1) = 1 \circ \varphi = 1$, $\varphi^*(f + g) = (f + g) \circ \varphi = f \circ \varphi + g \circ \varphi = \varphi^*(f) + \varphi^*(g)$, and $\varphi^*(fg) = (fg) \circ \varphi = (f \circ \varphi)(g \circ \varphi) = \varphi^*(f)\varphi^*(g)$.

Suppose, on the other hand, $\theta : k[W] \to k[V]$ is a k-homomorphism. Then, we can define a mapping $\psi : V \to W$ which is regular (hence, a morphism) and for which $\psi^* = \theta$. We do this as follows : if $W \subset A_k^m$, then $k[W] = k[Y_1, \ldots Y_m]/I(W)$ and let $\overline{y}_i$ be the residue class of Y_i. Then $\theta(\overline{y}_i) = f_i \in k[V]$ and f_i is a polynomial function on V. Let $\psi : V \to A_k^m$ be given by

$$\psi(v) = (f_1(v), \ldots, f_m(v)) \qquad (7.12)$$

for $v \in V$. We claim that $\psi(V) \subset W$. Since $W = V(I(W))$, it will be enough to show that if $v \in V$ and $g \in I(W)$, then $g(\psi(v)) = 0$. But

$g(\psi(v)) = g(f_1(v), \ldots, f_m(v)) = \theta(g(\overline{y}_1, \ldots, \overline{y}_m))(v) = 0$ since θ is a k-homomorphism and $g \in I(W)$. Thus, $\psi : V \to W$ and $\psi^*(h) = h \circ \psi = \theta(h)$ is in $k[V]$ if $h \in k[W]$ by the following :

Proposition 7.13. *A mapping $\phi : V \to W$ is regular if and only if $y_i \circ \phi$ is a regular function for each i, where $y_1, \ldots, y_m$ are the coordinate functions on $\mathbb{A}_k^m$.*

Proof: If ϕ is regular, then $y_i \circ \phi$ is a polynomial function by definition. If the $y_i \circ \phi$ are all regular functions, then, for any $f \in k[y_1, \ldots, y_m]$, $f \circ \phi = f(y_1 \circ \phi, \ldots, y_m \circ \phi)$ is a regular function.

Let us denote by $\mathrm{Hom}_k(V, W)$ the set of k-morphisms from V into W, by $\mathrm{Reg}(V, W)$ the set of regular maps of V into W, and by $\mathrm{Hom}_k(k[W], k[V])$ the set of k-homomorphisms of $k[W]$ into $k[V]$. We have shown that the natural map $\phi \to \phi^*$ of $\mathrm{Reg}(V, W)$ into $\mathrm{Hom}_k(k[W], k[V])$ is surjective. It is, in view of Proposition 7.13, also injective (since $\phi_1 \neq \phi_2$ implies there is a $v \in V$ with $\phi_1(v) \neq \phi_2(v)$ and hence, $(y_i \circ \phi_1)(v) \neq (y_i \circ \phi_2)(v)$ for some i so that $\phi_1^*(y_i) \neq \phi_2^*(y_i)$). In other words, we have :

Proposition 7.14. $\mathrm{Reg}(V, W) \simeq \mathrm{Hom}_k(k[W], k[V])$.

We shall next show that $\mathrm{Hom}(V, W)$ and $\mathrm{Hom}_k(k[W], k[V])$ also correspond naturally and so, for affine algebraic sets, morphisms and regular maps are the same thing.

A k-morphism $\phi : V \to W$ defines naturally a map $\phi^* : k[W] \to k[V]$ given by

$$\phi^*(g) = g \circ \phi \qquad (7.15)$$

This map is called the *comorphism* of ϕ. Again, ϕ^* is clearly a k-homomorphism, and we have a natural map $\phi \to \phi^*$ of $\mathrm{Hom}_k(V, W)$ into $\mathrm{Hom}_k(k[W], k[V])$.

Proposition 7.16. $\mathrm{Hom}_k(V, W) \simeq \mathrm{Hom}_k(k[W], k[V])$.

Proof: It will be enough to show that there is a map $\gamma : \mathrm{Hom}_k(k[W], k[V]) \to \mathrm{Hom}_k(V, W)$ such that $\gamma(h)^* = h$ and $\gamma(\phi^*) = \phi$, where $h \in \mathrm{Hom}_k(k[W], k[V])$ and $\phi \in \mathrm{Hom}_k(V, W)$. In view of theorem 7.5 and the fact that the map $\alpha \to \alpha \circ h$ carries $\mathrm{Hom}_k(k[V], k)$

into $\text{Hom}_k(k[W], k)$, we can define a map $\gamma(h) : V \to W$ by setting $\gamma(h)(v) = w$ where $\alpha_w = \alpha_v \circ h$ ($\alpha_v \in \text{Hom}_k(k[V], k)$ corresponding to v, etc.). If we show that $h(g) = g \circ \gamma(h) = \gamma(h)^*(g)$ for all $g \in k[W]$, then $\gamma(h)$ is a morphism and $\gamma(h)^* = h$. But $(g \circ \gamma(h))(v) = g(w) = \alpha_w(g) = (\alpha_v \circ h)(g) = \alpha_v(h(g)) = (h(g))(v)$. As for showing that $\gamma(\phi^*) = \phi$, we need only prove that $\alpha_{\gamma(\phi^*)(v)} = \alpha_{\phi(v)}$ for all v in V. But, if $g \in k[W]$, then $\alpha_{\gamma(\phi^*)(v)}(g) = \alpha_v(\phi^*(g)) = \alpha_v(g \circ \phi) = g(\phi(v)) = \alpha_{\phi(v)}(g)$.

We say that a k-morphism $\psi : V \to W$ is a *k-isomorphism* if it is bijective and ψ^{-1} is a k-morphism. We then have:

Corollary 7.17. *The morphism $\psi : V \to W$ is an isomorphism if and only if $\psi^* : k[W] \to k[V]$ is an isomorphism.*

Proposition 7.18. *Let $\psi : V \to W$ be a morphism and let $\psi^* : k[W] \to k[V]$ be the comorphism of ψ. Then: (i) ψ^* is injective if and only if $\overline{\psi(V)} = W$ i.e. $\psi(V)$ is dense in W; and, (ii) if ψ^* is surjective, then $\psi(V)$ is closed in W.*

Proof: We note that ψ^* is injective if and only if $\psi^*(f) = f \circ \psi = 0$ implies that $f = 0$. But this means that ψ^* is injective if and only if $I(\psi(V)) = 0$. Since $V(I(\psi(V))) = \overline{\psi(V)}$, (i) follows.

If ψ^* is surjective and $\mathfrak{a} = \ker \psi^*$, then $k[W]/\mathfrak{a}$ is isomorphic to $k[V]$. We claim that $\psi(V) = V(\mathfrak{a})$. If $w \in \psi(V)$ and $f \in \mathfrak{a}$, then $f(w) = f(\psi(v)) = \psi^*(f)(v) = 0$ where $w = \psi(v)$ so that $\psi(V) \subset V(\mathfrak{a})$. If $w \in V(\mathfrak{a})$, then the maximal ideal $\mathfrak{m}_w \supset \mathfrak{a}$ and there is a maximal ideal $\mathfrak{m}_v$ in $k[V]$ such that $\overline{\mathfrak{m}}_w = \mathfrak{m}_w/\mathfrak{a}$ "=" $\mathfrak{m}_v$ (as $k[W]/\mathfrak{a} \simeq k[V]$). But then $\psi(v) = w$ so that $w \in \psi(V)$. Thus, (ii) is established.

Example 7.19.   Let $V = \mathbb{A}_k^1$ and $W \subset \mathbb{A}_k^2$ be given by $W = \{(\xi_1, \xi_2); \xi_2 = \xi_1^2\}$. Then the map $\phi : V \to W$ given by $\phi(\xi) = (\xi, \xi^2)$ is an isomorphism of V and W since $k[W] \simeq k[\mathbb{A}_k^1]$ (cf. example 7.4).

Example 7.20.   (cf. example 4.11) Let $L : \mathbb{A}_k^2 \to \mathbb{A}_k^2$ be given by $L(x, y) = (x, -xy)$. L is a morphism. The algebraic set $V(I(L(\mathbb{A}_k^2)))$ is the closure of $L(\mathbb{A}_k^2)$. Since 0 is the only polynomial which vanishes on $L(\mathbb{A}_k^2)$ i.e. $I(L(\mathbb{A}_k^2)) = (0)$, the closure of $L(\mathbb{A}_k^2)$ is all of $\mathbb{A}_k^2$. However, $L(\mathbb{A}_k^2)$ does not contain the points $(0, \xi)$, $\xi \neq 0$ and so, the image under a morphism need not be closed.

Example 7.21.   Let $L : \mathbb{A}_k^{2n} \to \mathbb{A}_k^{2n}$ be given by $L(\xi_0, \ldots, \xi_{n-1},$
$\eta_0, \ldots, \eta_{n-1}) = (c(\xi)b, c(\xi)A(\eta)b, \ldots, c(\xi)A(\eta)^{2n-1}b)$ where

$$
A(\eta) = \begin{bmatrix}
0 & 1 & 0 & \cdots & 0 \\
0 & 0 & 1 & \cdots & 0 \\
\vdots & \vdots & & & \vdots \\
\vdots & \vdots & & & 1 \\
-\eta_0 & -\eta_1 & \cdots & \cdots & -\eta_{n-1}
\end{bmatrix},
$$

$$
b = \begin{bmatrix} 0 \\ 0 \\ \vdots \\ 1 \end{bmatrix}, c(\xi) = [\xi_0, \ldots, \xi_{n-1}]
$$

(L is just the Laurent map "extended" to all of $\mathbb{A}_k^{2n}$). Then L is a morphism.

Example 7.22.   Let $\phi : \mathbb{A}_k^{n^2+2n} \to \mathbb{A}_k^{2n}$ be given by $\phi(X, Y, Z) = (ZY, ZXY, \ldots, ZX^{2n-1}Y)$. Then ϕ is a morphism. For instance, if $n = 2$, then $\phi(X_{11}, X_{12}, X_{21}, X_{22}, Y_1, Y_2, Z_1, Z_2) = (Z_1Y_1 + Z_2Y_2, Z_1X_{11}Y_1 + Z_1X_{12}Y_2 + Z_2X_{21}Y_1 + Z_2X_{22}Y_{21}, \ldots)$.

Example 7.23.   Let $V = \mathbb{A}_k^1$ and $W \subset \mathbb{A}_k^2$ be given by $W = \{(\xi_1, \xi_2) : \xi_1^3 - \xi_2^2 = 0\}$. The map $\phi : V \to W$ with $\phi(\xi) = (\xi^2, \xi^3)$ is a morphism. Since $\xi \neq \xi'$ implies $\xi^2 \neq \xi'^2$ or $\xi^3 \neq \xi'^3$, ϕ is injective. Obviously, $\phi(0) = (0, 0) \in W$. If $(\xi_1, \xi_2) \neq (0, 0)$ is an element of W, then, letting $\xi = \xi_2/\xi_1$, we have $\phi(\xi) = (\xi_2^2/\xi_1^2, \xi_2^3/\xi_1^3) = (\xi_1^3/\xi_1^2, \xi_2\xi_1^3/\xi_1^3) = (\xi_1, \xi_2)$ so that ϕ is surjective. But $k[W] = k[X^2, X^3]$ (since $k[W] = k[Y_1, Y_2]/(Y_1^3 - Y_2^2) \simeq k[X^2, X^3]$ under the map which sends $\overline{Y}_1 \to X^2$, $\overline{Y}_2 \to X^3$) and so ϕ is not an isomorphism. (In other words, a bijective morphism need not be an isomorphism.)

The fact that morphisms of affine algebraic sets are restrictions of morphisms of the ambient affine spaces (i.e. of polynomial maps) is special and will not be true in general. However, it makes the continuity of morphisms in the Zariski topology almost obvious.

Proposition 7.24.   *A regular mapping* $\phi : V \to W$ *is continuous.*

Proof: If W_1 is a closed subset of W, then $W_1 = \tilde{W}_1 \cap W$ where $\tilde{W}_1$ is closed in $\mathbb{A}_k^m$ and so, W_1 is closed in $\mathbb{A}_k^m$. Let ϕ be the restriction of the polynomial map $F = (F_1, \ldots, F_m)$ of $\mathbb{A}_k^N$ into $\mathbb{A}_k^m$ where $F_i \in k[x_1, \ldots, x_N]$. Then $\phi^{-1}(W_1) = F^{-1}(W_1) \cap V$ and so, $\phi^{-1}(W_1)$ is closed by the following lemma.

Lemma 7.25. *If $F : \mathbb{A}_k^N \to \mathbb{A}_k^m$ is a polynomial mapping, then F is continuous.*

Proof: Let $W \subset \mathbb{A}_k^m$ be an affine algebraic set and let $G_1, \ldots, G_r$ be a basis of $I(W)$. Then $V = \{\xi \in \mathbb{A}_k^N : (G_i \circ F)(\xi) = 0, i = 1, \ldots, r\}$ is an affine algebraic set. Moreover, if $\xi \in V$, then $F(\xi) \in W$ and conversely so that $V = F^{-1}(W)$.

8. The Laurent Isomorphism Theorem

Now, one of our goals is to show that the Laurent map $L : \text{Rat}(n, k) \to \text{Hank}(n, k)$ is a k-isomorphism. However, $\text{Rat}(n, k)$ and $\text{Hank}(n, k)$ are not affine algebraic sets but rather are irreducible principal open sets. Thus, we must eventually extend the concepts of regularity and morphism. We begin with:

Proposition 8.1. *Let V_f be a principal affine open subset of $V \subset \mathbb{A}_k^N$. Then there is a natural bijection ψ_f between V_f and an affine algebraic set in $\mathbb{A}_k^{N+1}$.*

Proof: Let $f_1(x_1, \ldots, x_N), \ldots, f_r(x_1, \ldots, x_N)$ be a basis of $I(V)$ and let $h(x_1, \ldots, x_N, x_{N+1}) = 1 - x_{N+1}f(x_1, \ldots, x_N)$. Let $W \subset \mathbb{A}_k^{N+1}$ be the zero set of the f_i and h i.e. $W = V((f_1, \ldots, f_r, h))$. We define the map ψ_f by setting

$$\psi_f(\xi_1, \ldots, \xi_N) = (\xi_1, \ldots, \xi_N, 1/f(\xi_1, \ldots, \xi_N)) \tag{8.2}$$

for $\xi = (\xi_1, \ldots, \xi_N) \in V_f$. Clearly ψ_f is injective. If $\eta = (\eta_1, \ldots, \eta_N, \eta_{N+1}) \in W$, then $f_i(\eta_1, \ldots, \eta_N) = 0$, $i = 1, \ldots, r$ and $h(\eta) = 1 - \eta_{N+1}f(\eta_1, \ldots, \eta_N) = 0$ together imply that $(\eta_1, \ldots, \eta_N) \in V_f$. Thus ψ_f is surjective.

Recalling that $\text{Rat}(n, k) = (\mathbb{A}_k^{2n})_\rho$ where ρ is the resultant and that $\text{Hank}(n, k) = (\mathbb{A}_k^{2n})_\theta$ where $\theta = \det H$, we may view $\text{Rat}(n, k)$ as the algebraic set $V((1 - x_{2n+1}\rho))$ in $\mathbb{A}_k^{2n+1}$ and $\text{Hank}(n, k)$ as the algebraic set $V((1 - x_{2n+1}\det H))$ in $\mathbb{A}_k^{2n+1}$. We shall eventually extend the notion of a morphism in such a way that the mappings $\psi_\rho : \text{Rat}(n, k) \to V((1 - x_{2n+1}\rho))$ and $\psi_\theta : \text{Hank}(n, k) \to V((1 - x_{2n+1}\det H))$ are k-isomorphisms. For the moment, however, we shall adopt the point of view that $\text{Rat}(n, k) = V((1 - x_{2n+1}\rho))$ and $\text{Hank}(n, k) = V((1 - x_{2n+1}\theta))$. From this point of view, we can describe the Laurent map L as follows: let $(x_1, \ldots, x_{2n}, x_{2n+1})$ and $(y_1, \ldots, y_{2n}, y_{2n+1})$ be coordinate functions on $\mathbb{A}_k^{2n+1}$ and define a polynomial map $\tilde{L} : \mathbb{A}_k^{2n+1} \to \mathbb{A}_k^{2n+1}$ by setting

$$(y_j \circ \tilde{L})(x_1, \ldots, x_{2n}, x_{2n+1}) = \tilde{c}(x_1, \ldots, x_{2n+1})\tilde{A}^{j-1}(x_1, \ldots, x_{2n+1})\tilde{b} \tag{8.3}$$

for $j = 1, 2, \ldots 2n$ and

$$(y_{2n+1} \circ \tilde{L})(x_1, \ldots, x_{2n}, x_{2n+1}) = -x_{2n+1} \tag{8.4}$$

for $n \geq 2*$, where

$$\tilde{A}(x) = \begin{bmatrix} 0 & 1 & 0 & \cdots & 0 \\ 0 & 0 & 1 & \cdots & 0 \\ \vdots & \vdots & & & \vdots \\ \vdots & \vdots & & & 1 \\ -x_{n+1} & -x_{n+2} & \cdots & \cdots & -x_{2n} \end{bmatrix}, \tag{8.5}$$

$$\tilde{b} = \begin{bmatrix} 0 \\ 0 \\ \vdots \\ 1 \end{bmatrix}, \tilde{c}(x) = [x_1, \ldots, x_n]$$

(cf. (2.9)). Then L is simply the restriction of $\tilde{L}$ to Rat $(n, k) = V((1 - x_{2n+1}\rho))$. Thus, L is a morphism and hence, is continuous. L is surjective by Hankel's Theorem (2.5).

We can also study the map $F : \text{Hank}(n, k) \to \text{Rat}(n, k)$ from the same point of view. Again, we let $(y_1, \ldots, y_{2n}, y_{2n+1})$ and $(x_1, \ldots, x_{2n}, x_{2n+1})$ be coordinate functions on $\mathbb{A}_k^{2n+1}$ and we define a polynomial map $\tilde{F} : \mathbb{A}_k^{2n+1} \to \mathbb{A}_k^{2n+1}$ by setting

$$(x_s \circ \tilde{F})(y_1, \ldots, y_{2n}, y_{2n+1}) = y_{n-s+1} - \sum_{j=1}^{n-s}(\mathbf{Y}^+)^{s+j} \begin{bmatrix} y_{n+1} \\ \vdots \\ y_{2n} \end{bmatrix} y_j y_{2n+1} \tag{8.6}$$

for $s = 1, \ldots, n - 1$ and

$$(x_n \circ \tilde{F})(y_1, \ldots, y_{2n}, y_{2n+1}) = y_1 \tag{8.7}$$

and

$$(x_{n+t} \circ \tilde{F})(y_1, \ldots, y_{2n}, y_{2n+1}) = (-\mathbf{Y}^+)^t \begin{bmatrix} y_{n+1} \\ \vdots \\ y_{2n} \end{bmatrix} y_{2n+1} \tag{8.8}$$

---

* For $n = 1$, $y_3 \circ \tilde{L} = x_3$.

for $t = 1, \ldots, n$ and

$$(x_{2n+1} \circ \tilde{F})(y_1, \ldots, y_{2n}, y_{2n+1}) = -y_{2n+1} \qquad (8.9)$$

for $n \geq 2$ *,(cf. (2.14)) where Y^+ is the formal adjoint of the matrix

$$Y = \begin{bmatrix} y_1 & y_2 & \cdots & y_n \\ y_2 & y_3 & \cdots & y_{n+1} \\ \vdots & \vdots & & \vdots \\ y_n & y_{n+1} & \cdots & y_{2n-1} \end{bmatrix} \qquad (8.10)$$

and Y^t is its t-th row. Then F is simply the restriction of $\tilde{F}$ to Hank $(n, k) = V((1 - y_{2n+1} \det Y))$. Thus, F is also a morphism.

Example 8.11.   Let $n = 2$ and let $(x_1, \ldots, x_4, x_5)$ and $(y_1, \ldots, y_4, y_5)$ be coordinates on $\mathbb{A}_k^5$. Then $\mathrm{Rat}(2, k) = V((1 - x_5(x_1^2 - x_1 x_2 x_4 + x_2^2 x_3)))$ and Hank $(2, k) = V((1 - y_5(y_1 y_3 - y_2^2)))$. The maps $\tilde{L}$ and $\tilde{F}$ are given by

$$y_1 \circ \tilde{L} = x_2$$
$$y_2 \circ \tilde{L} = x_1 - x_2 x_4$$
$$y_3 \circ \tilde{L} = -x_1 x_4 - x_2 x_3 + x_2 x_4^2$$
$$y_4 \circ \tilde{L} = -x_1 x_3 + x_1 x_4^2 + 2x_2 x_3 x_4 - x_2 x_4^3$$
$$y_5 \circ \tilde{L} = -x_5$$

$$x_1 \circ \tilde{F} = (2y_1 y_2 y_3 - y_1^2 y_4 - y_2^3)y_5$$
$$x_2 \circ \tilde{F} = y_1$$
$$x_3 \circ \tilde{F} = (y_2 y_4 - y_3^2)y_5$$
$$x_4 \circ \tilde{F} = (-y_1 y_4 + y_2 y_3)y_5$$
$$x_5 \circ \tilde{F} = -y_5$$

and L and F are their restrictions to $\mathrm{Rat}(2, k)$ and Hank$(2, k)$ respectively. We observe that $(y_1 \circ L)(y_3 \circ L) - (y_2 \circ L)^2 = -x_1 x_2 x_4 - x_2^2 x_3 + x_2^2 x_4^2 - x_1^2 + 2x_1 x_2 x_4 - x_2^2 x_4^2 = -x_1^2 + x_1 x_2 x_4 - x_2^2 x_3$ so that $\{(y_1 \circ L)(y_3 \circ L) - (y_2 \circ L)^2\}(y_5 \circ L) = x_5(x_1^2 - x_1 x_2 x_4 + x_2^2 x_3) = 1$ i.e. L maps $\mathrm{Rat}(2, k)$ into Hank$(2, k)$. Similarly, F maps Hank$(2, k)$ into $\mathrm{Rat}(2, k)$. Let us examine

---

* For $n = 1$, $x_3 \circ \tilde{F} = y_3$.

the map $F \circ L$. We have $x_1 \circ (F \circ L) = \{2x_2(x_1 - x_2x_4)(-x_1x_4 - x_2x_3 + x_2x_4^2) - x_2^2(-x_1x_3 + x_1x_4^2 + 2x_2x_3x_4 - x_2x_4^3) - (x_1 - x_2x_4)^3\} (-x_5) = x_1(x_1^2 - x_1x_2x_4 + x_2^2x_3)x_5 = x_1$, $x_2 \circ (F \circ L) = x_2$, $x_3 \circ (F \circ L) = \{(x_1 - x_2x_4)(-x_1x_3 + x_1x_4^2 + 2x_2x_3x_4 - x_2x_4^3) - (-x_1x_4 - x_2x_3 + x_2x_4^2)^2\}(-x_5) = x_3$, $x_4 \circ (F \circ L) = \{(-x_2)(-x_1x_3 + x_1x_4^2 + 2x_2x_3x_4 - x_2x_4^3) + (x_1 - x_2x_4)(-x_1x_4 - x_2x_3 + x_2x_4^2)\}(-x_5) = x_4(x_1^2 + x_2^2x_3 - x_1x_2x_4)x_5 = x_4$, and $x_5 \circ (F \circ L) = x_5$ so that $F \circ L$ is the identity on Rat$(2, k)$. Similarly, $L \circ F$ is the identity on Hank $(2, k)$. Thus, L is a k-isomorphism.

If we examine what we have done so far, we can see that the Laurent map L is a surjective morphism and that the map F is a morphism. If we can show that L is injective and that $F = L^{-1}$, then we will have demonstrated the following :

Theorem 8.12.   *The Laurent map L is a k-isomorphism between* Rat $(n, k) = V((1 - x_{2n+1}\rho))$ *and* Hank$(n, k) = V((1 - x_{2n+1}\theta))$.

In other words, the notions (a), (b), (c) of a scalar linear system are essentially the same.

Lemma 8.13. *Let $u_1, \ldots, u_n$ be variables and let*

$$A_u = \begin{bmatrix} 0 & 1 & 0 & \cdots & 0 \\ 0 & 0 & 1 & \cdots & 0 \\ \vdots & \vdots & & & 1 \\ -u_1 & -u_2 & \cdots & \cdots & -u_n \end{bmatrix}$$

$$g_u = \begin{bmatrix} u_2 & u_3 & \cdots & u_n & 1 \\ u_3 & u_4 & \cdots & 1 & 0 \\ \vdots & \vdots & & & \vdots \\ & & 1 & & \vdots \\ 1 & 0 & \cdots & \cdots & 0 \end{bmatrix}$$

$$\widehat{A}_u = \begin{bmatrix} 0 & 0 & \cdots & 0 & -u_1 \\ 1 & 0 & & \vdots & -u_2 \\ 0 & 1 & & \vdots & \vdots \\ \vdots & \vdots & & \vdots & \vdots \\ 0 & 0 & \cdots & 1 & -u_n \end{bmatrix}$$

Then $g_u A_u = \widehat{A}_u g_u$ *i.e.* $g_u A_u g_u^{-1} = \widehat{A}_u$.

Proof: Simply compute.

Lemma 8.14. *Let L be the Laurent map and let F be the associated map of Hank (n,k) into Rat (n,k). Then L∘F is the identity on Hank* (n, k).

Proof: Let h_i be the restriction of y_i to Hank(n, k) for $i = 1, \ldots, 2n$ and let $\boldsymbol{H} = (h_{i+j-1})_{i,j=1}^n$ be the restriction of $\boldsymbol{Y}$ to Hank (n, k). Then $1/y_{2n+1} = \det \boldsymbol{H}$ on Hank (n, k) and we may write the equations of F as follows:

$$\boldsymbol{H} \begin{bmatrix} -x_{n+1} \circ F \\ \vdots \\ -x_{2n} \circ F \end{bmatrix} = \begin{bmatrix} h_{n+1} \\ \vdots \\ h_{2n} \end{bmatrix} \tag{8.15}$$

$$[x_1 \circ F \cdots x_n \circ F] = [h_1 \cdots h_n]g_F$$

where g_F is of the form g_u of lemma 8.13 with $u_1 = +x_{n+1} \circ F, \ldots, u_n = +x_{2n} \circ F$. What we must then show is that

$$\tilde{c}(x \circ F)\widetilde{A}^{j-1}(x \circ F)\tilde{b} = h_j \tag{8.16}$$

for $j = 1, \ldots, 2n$ where $\tilde{c}(x \circ F)$, $\tilde{A}(x \circ F)$, and $\tilde{b}$ are given by (8.5). However, in view of the previous lemma and the fact that g_F is a nonsingular $n \times n$-matrix, we have

$$\tilde{c}(x \circ F)\tilde{A}^{j-1}(x \circ F)\tilde{b} = (\tilde{c}(x \circ F)g_F^{-1})(g_F \tilde{A}^{j-1} g_F^{-1})g_F \tilde{b}$$

$$= [h_1 \cdots h_n]\widehat{A}_F^{j-1} \begin{bmatrix} 1 \\ 0 \\ \vdots \\ 0 \end{bmatrix} \tag{8.17}$$

where $\widehat{A}_F$ is of the form $\widehat{A}_u$ of lemma 8.13 with $u_1 = +x_{n+1} \circ F, \ldots, u_n = +x_{2n} \circ F$. However, it follows from (8.17) and the form of $\widehat{A}_F$ and the Cayley-Hamilton Theorem and (8.15) that equation (8.16) holds for $j = 1, \ldots, 2n$.

Lemma 8.18. *Let* $u_1, \ldots, u_n$ *be variables and let* A_u, g_u *be of the form given in lemma 8.3. If* ϵ_n *is the unit column vector with 1 in the n-th row, then* $[\epsilon_n \ A_u \epsilon_n \cdots A_u^{n-1} \epsilon_n] = g_u^{-1}$.

Proof: Simply compute $g_u[\epsilon_n \; A_u \epsilon_n \cdots A_u^{n-1} \epsilon_n](= I_n)$.

Example 8.19. Let $n = 4$. Then lemma 8.18 states that

$$
\begin{bmatrix} u_2 & u_3 & u_4 & 1 \\ u_3 & u_4 & 1 & 0 \\ u_4 & 1 & 0 & 0 \\ 1 & 0 & 0 & 0 \end{bmatrix}
\begin{bmatrix} 0 & 0 & 0 & 1 \\ 0 & 0 & 1 & -u_4 \\ 0 & 1 & -u_4 & -u_3 + u_4^2 \\ 1 & -u_4 & -u_3 + u_4^2 & -u_2 + 2u_3 u_4 - u_4^3 \end{bmatrix}
$$
$$
= \begin{bmatrix} 1 & 0 & 0 & 0 \\ 0 & 1 & 0 & 0 \\ 0 & 0 & 1 & 0 \\ 0 & 0 & 0 & 1 \end{bmatrix}
$$

Lemma 8.20. *If L is the Laurent map and F is the associated map of Hank (n, k) into Rat (n, k). Then $F \circ L$ is the identity on Rat (n, k).*

Proof: We must show that $x_j \circ F \circ L = x_j$ on Rat (n, k). Now, for $j = 1, \ldots, n$, we have

$$[x_1 \circ (F \circ L) \cdots x_n \circ (F \circ L)] = [\tilde{c}\tilde{b} \cdots \tilde{c}\tilde{A}^{n-1}\tilde{b}]g_F$$

where g_F is of the form g_u of lemma 8.13 with $u_j = x_{n+j} \circ (F \circ L)$, $j = 1, \ldots, n$. Suppose for the moment that $x_{n+j} \circ (F \circ L) = x_j$ for $j = 1, \ldots, n$. It will then follow from lemma 8.18 that

$$[x_1 \circ (F \circ L) \cdots x_n \circ (F \circ L)] = [x_1 \cdots x_n][\epsilon_n \tilde{A} \epsilon_n \cdots \tilde{A}^{n-1} \epsilon_n]g_F$$
$$= [x_1 \cdots x_n]$$

i.e. $x_j \circ (F \circ L) = x_j$ for $j = 1, \ldots, n$. Since L maps Rat (n, k) into Hank (n, k), we have $[\tilde{c}\tilde{A}^{n+j}\tilde{b}]_{j=1}^n = \boldsymbol{H}(x)[-x_{n+j}]_{j=1}^n$ where $\boldsymbol{H}(x) = [\tilde{c}\tilde{A}^{i+j-1}\tilde{b}]_{i,j=1}^n$. But $[x_{n+j} \circ (F \circ L)]_{j=1}^n = \boldsymbol{H}(x)^{-1}[\tilde{c}\tilde{A}^{n+j}\tilde{b}]_{j=1}^n$ and so we have $x_{n+j} \circ (F \circ L) = x_{n+j}$ for $j = 1, \ldots, n$. Thus, the lemma is established.

In view of lemmas (8.14) and (8.20), we have established the Laurent Isomorphism Theorem (8.12). To fully complete the proof, we shall have to extend the ideas in such a way that the mappings ψ_ρ and ψ_θ are k-isomorphisms. This we shall do in the next section.

We note that the rationale for our proof of the Laurent Isomorphism Theorem is based on the structure of the state space representation and on the relation of the state space representation to the other representations. The fact that lemmas 8.13 and 8.18 can be done "globally" is special to the scalar situation.

9. Affine Algebraic Geometry: Regular Functions and Morphisms II

We now turn our attention to the issue of generalizing the notions of regular function and morphism. We recall that a complex meromorphic function is said to be *regular at a point* if the point is not a pole of the function. Similarly, a complex meromorphic function is said to be *regular on a set* if it has no poles in the set.

Let V be an affine variety (i.e. V is irreducible) so that $k[V]$ is an integral domain and $\mathfrak{p} = I(V)$ is a prime ideal. We denote the quotient field of $k[V]$ by $k(V)$ and call it the *function field of* V. If ξ is a point of V and $\mathfrak{m}_\xi = \{f : f \in k[V], f(\xi) = 0\}$, then $\mathfrak{m}_\xi$ is a maximal ideal in $k[V]$ and $\mathfrak{m}_\xi = M_\xi/\mathfrak{p}$ where $M_\xi = (x_1 - \xi_1, \ldots, x_N - \xi_N)$ is the maximal ideal in the polynomial ring $k[X_1, \ldots, X_N]$ consisting of those polynomials which vanish at ξ.

Definition 9.1.　The set $\mathcal{O}_{\xi,V} = \{f/g : f, g \in k[V], g(\xi) \neq 0\}$ is called the *local ring of* V *at* ξ.

We observe that $\mathcal{O}_{\xi,V}$ is indeed a local ring (i.e. a ring with exactly one maximal ideal) since the set of functions f/g with $(f/g)(\xi) = 0$ forms a maximal ideal and if $(f/g)(\xi) \neq 0$, then g/f is an element of $\mathcal{O}_{\xi,V}$ (i.e. every element of $\mathcal{O}_{\xi,V}$ which does not vanish at ξ is invertible). We also observe that $\mathcal{O}_{\xi,V}$ is an integral domain with the same quotient field as $k[V]$, namely the function field $k(V)$ of V.

Now, $1 \notin \mathfrak{m}_\xi$ and if $f_1, f_2 \notin \mathfrak{m}_\xi$, then $f_1 f_2 \notin \mathfrak{m}_\xi$ so that $\mathcal{O}_{\xi,V} = \{f/g : g \notin \mathfrak{m}_\xi\}$. We have:

Definition 9.2.　Let R be a ring. A subset M of R is *multiplicatively closed* if $1 \in M$, $0 \notin M$ and $m, m' \in M$ implies $mm' \in M$.

Some examples of multiplicatively closed sets are $M_f = \{f^n : f \neq 0, n \geq 0\}$ and $M_\mathfrak{p} = R - \mathfrak{p} = \{r : r \notin \mathfrak{p}\}$ where $\mathfrak{p}$ is a prime ideal. If M is a multiplicatively closed subset of R, then we can define an equivalence relation on $R \times M$ as follows: $(r, m) \sim (r', m')$ if there is an m'' such that $m''(m'r - mr') = 0$. We denote the set of equivalence classes by R_M and we let r/m denote the equivalence class of (r, m). By defining addition and multiplication of these fractions r/m in the usual way i.e.

$$(r/m) + (r'/m') = (m'r + mr')/(mm')$$
$$(r/m)(r'/m') = (rr'/mm')$$

we make R_M into a ring.

Definition 9.3. R_M is called the *quotient ring of R with respect to M* or the *ring of fractions of R with respect to M*.

If $M = M_f = \{f^n : f \neq 0, n \geq 0\}$, then we write R_f in place of R_{M_f} and if $M = M_{\mathfrak{p}} = \{r : r \notin \mathfrak{p}\}$ where $\mathfrak{p}$ is a prime ideal, then we wirte $R_{\mathfrak{p}}$ in place of $R_{M_{\mathfrak{p}}}$.

Proposition 9.4.   *If R is an integral domain, then so is R_M.*

Proof: If $(r/m)(r'/m') = 0$ then $rr' = 0$ and so r or r' is 0.

Proposition 9.5.   *If $\mathfrak{p}$ is a prime ideal, then $R_{\mathfrak{p}}$ is a local ring.*

Proof: The elements r/m with $r \in \mathfrak{p}$ form an ideal $\mathfrak{m}_{\mathfrak{p}} = \mathfrak{p}R_{\mathfrak{p}}$ in $R_{\mathfrak{p}}$. If r_1/m_1 is not in $\mathfrak{m}_{\mathfrak{p}}$, then r_1 is not in $\mathfrak{p}$ and m_1/r_1 is in $R_{\mathfrak{p}}$ i.e. r_1/m_1 is a unit in $R_{\mathfrak{p}}$. Thus $\mathfrak{m}_{\mathfrak{p}}$ is the unique maximal ideal in $R_{\mathfrak{p}}$.

Proposition 9.6.   *If R is an integral domain, then the natural homomorphism $r \longrightarrow r/1$ of R into R_M is an isomorphism.*

Proof: It is enough to show that the map is injective. But $r/1 - r'/1 = 0$ means that $m(r - r') = 0$ for some $m(\neq 0)$ in M. Since R is an integral domain, $r - r' = 0$.

We now note that

$$\mathcal{O}_{\xi,V} = k[V]_{\mathfrak{m}_{\xi}} = (k[X_1, \ldots, X_N]_{M_{\xi}})/\mathfrak{p}k[X_1, \ldots, X_N]_{M_{\xi}} \qquad (9.7)$$

where $\mathfrak{p} = I(V)$. In other words, $\mathcal{O}_{\xi,V}$ is the quotient ring of $k[V]$ with respect to the multiplicative set $k[V] - \mathfrak{m}_{\xi} = \{f : f \notin \mathfrak{m}_{\xi}\}$.

Definition 9.8.   Let V be an affine variety and let $\xi \in V$. Call a function $f : V \longrightarrow \mathbb{A}_k^1$ *regular at* ξ is there is an open neighborhood U (in V) of ξ and polynomials F_1, F_2 in $k[X_1, \ldots, X_N]$ such that F_2 does not vanish at any point of U and $f(u) = F_1(u)/F_2(u)$ for all u in U. A function is *regular on* $W \subset V$ if it is regular at each point of W.

Proposition 9.9.   *A function* $f : V \longrightarrow \mathbb{A}_k^1$ *is regular at* ξ *if and only if* $f \in \mathcal{O}_{\xi,v}$.

Proof: If $f \in \mathcal{O}_{\xi,V}$, then $f = g/h$ with $h(\xi) \neq 0$. Let G and H be polynomials whose restrictions to V are g, h respectively. If $U = V_h$ is the principal open subset of V defined by h, then $\xi \in U \subset V$, $H(\xi) = h(\xi) \neq 0$, and $f(u) = g(u)/h(u) = G(u)/H(u)$ for all u in U. Conversely, if f is regular at ξ, then $f = F_1/F_2$, $F_2 \neq 0$ at ξ. Letting f_1 and f_2 be the restrictions of F_1 and F_2, respectively, to V, we have $f = f_1/f_2$ with $f_2(\xi) \neq 0$ so that $f \in \mathcal{O}_{\xi,V}$.

This immediately leads to the following:

Definition 9.10.   If U is open in V, then $\mathcal{O}_V(U) = \cap_{\xi \in U} \mathcal{O}_{\xi,V}$ is the *ring of regular functions on* U.

Proposition 9.11. $k[V] = \mathcal{O}_V(V) = \cap_{\xi \in V} \mathcal{O}_{\xi,V}$ *(i.e. the local rings* $\mathcal{O}_{\xi,V}$ *determine* $k[V]$ *and the notion of regularity in definition 9.8 extends our previous definition.)*

Proof ([M-2]): Clearly $k[V] \subset \mathcal{O}_{\xi,V}$. So suppose that $f \in \mathcal{O}_{\xi,V}$ for all $\xi \in V$. Let $\mathfrak{a}_f = \{g \in k[X_1, \ldots, X_N] : \bar{g}f \in k[V]$ where $\bar{g} = g \bmod \mathfrak{p}\}$. Then $\mathfrak{a}_f$ is an ideal in $k[X_1, \ldots, X_N]$. Since $f \in \mathcal{O}_{\xi,V}$ $f = F_\xi/G_\xi$ where $F_\xi, G_\xi \in k[X_1, \ldots, X_N]$ and $G_\xi(\xi) \neq 0$. Hence, $G_\xi \in \mathfrak{a}_f$ and $\xi \notin V(\mathfrak{a}_f)$ for all $\xi \in V$. But $\mathfrak{p} = I(V) \subset \mathfrak{a}_f$ implies $V(\mathfrak{a}_f) \subset V = V(\mathfrak{p})$ so that $V(\mathfrak{a}_f) = \phi$. However, $1 \in \mathfrak{a}_f$ by the Nullstellensatz and so, $f \in k[V]$.

Proposition 9.12. $\mathcal{O}_V(V_f) = k[V]_f$ *(i.e. the ring of regular functions on* V_f *is the quotient ring of* $k[V]$ *with respect to the multiplicatively closed set* $M_f = \{f^n : n \geq 0\}$).

Proof: If h/f^n is an element of $k[V]_f$, then $h/f^n \in \mathcal{O}_{\xi,V}$ for $\xi \in V_f$.

Conversely, if $h \in \mathcal{O}_V(V_f)$ and if $\mathfrak{a}_h = \{g : gh \in k[V]\}$, then $\mathfrak{a}_h$ is an ideal in $k[V]$. We claim that $V(\mathfrak{a}_h) \subset V((f))$. If $\xi \in V_f$, then $h = h_1/h_2$ with $h_2(\xi) \neq 0$. But $h_2 \in \mathfrak{a}_h$ and so $\xi \notin V(\mathfrak{a}_h)$. However, $V(\mathfrak{a}_h) \subset V((f))$ implies that $f \in \sqrt{\mathfrak{a}_h}$. It follows that $f^n h \in k[V]$ and, hence, that $h \in k[V]_f$.

We are now ready to extend the notion of a morphism. In particular, we have:

Definition 9.13.   Let V and W be quasi-affine varieties. A mapping $\phi : V \to W$ is a *morphism* (or *k-morphism*) if (i) ϕ is continuous and (ii) $\phi^*(g) = g \circ \phi$ is a regular function on V if g is a regular function on W. We also say that a k-morphism $\psi : V \to W$ is a *k-isomorphism* if it is bijective and ψ^{-1} is a k-morphism.

Proposition 9.14.   *Let V_f be a principal affine open subset of the variety $V \subset \mathbb{A}_k^N$ and let ψ_f be the natural bijection between V_f and the affine algebraic set W in $\mathbb{A}_k^{N+1}$ (see proposition 8.1). Then ψ_f is a k-isomorphism.*

Proof: Let $F(X_1, \ldots, X_N) \in k[X_1, \ldots, X_N]$ such that F restricted to V is f. If g is an element of $k[W]$ and $G(X_1, \ldots, X_N, X_{N+1}) = \sum_{i=0}^{n} G_i(X_1, \ldots, X_N) X_{N+1}^i$ is a polynomial whose restriction to W is g, then $\psi_f^*(g) = g \circ \psi_f = H(X_1, \ldots, X_N)/F^n(X_1, \ldots, X_N)$ restricted to V_f for some polynomial H and so, $\psi_f^*(g) \in \mathcal{O}_V(V_f) = k[V]_f$. In other words, ψ_f is a morphism. But we have $k[W] = k[X_1, \ldots, X_N, X_{N+1}]/(I(V), 1 - X_{N+1}F)$ which is k-isomorphic to $R = k[V][X_{N+1}]/(1 - fX_{N+1})k[V][X_{N+1}]$. So, it will be enough to show that R is k-isomorphic to $k[V]_f$. We define first a k-homomorphism $\alpha : k[V][X_{N+1}] \to k[V]_f$ by setting α equal to the identity on $k[V]$ and $\alpha(X_{N+1}) = 1/f$. Clearly $\alpha(1 - fX_{N+1}) = 0$ and so we can extend α to a k-homomorphism $\tilde{\alpha} : R \to k[V]_f$. Since every element of $k[V]_f$ is of the form g/f^n, $\tilde{\alpha}$ is surjective and in fact, every element of $k[V]_f$ is of the form $\alpha(gX_{N+1}^n)$ where $g \in k[V]$. If $\alpha(gX_{N+1}^n) = g/f^n = 0$, then, in view of proposition 9.4, $g = 0$ and so, $\tilde{\alpha}$ is injective.

Corollary 9.15.   *The Laurent map $L : \mathrm{Rat}(n, k) \to \mathrm{Hank}(n, k)$ is a k-isomorphism.*

Proof: The maps ψ_ρ and ψ_θ are k-isomorphisms.

Example 9.16.   (cf. example 6.9).   Now let $GL(2,k) = \{g : g \in M(2,2;k),\ \det g \neq 0\} = (\mathbb{A}_k^4)_\Delta$ where $\Delta(X) = \Delta(X_{11}, X_{12}, X_{21}, X_{22}) = X_{11}X_{22} - X_{12}X_{21}$. Let $\beta : G \to G$ be given by $\beta(g) = g^{-1}$ i.e. if

$$g = \begin{bmatrix} g_{11} & g_{12} \\ g_{21} & g_{22} \end{bmatrix}$$

then

$$\beta(g) = \begin{bmatrix} g_{22} & -g_{12} \\ -g_{21} & g_{11} \end{bmatrix} / (g_{11}g_{22} - g_{21}g_{12})$$

Then β is a k-isomorphism of $GL(2,k)$ onto $GL(2,k)$.   Clearly β is bijective. Observing that $\mathcal{O}_{\mathbb{A}_k^4}(GL(2,k)) = k[\mathbb{A}_k^4]_\Delta$ and that $\beta^*(X_{11}) = X_{22}/\Delta$, $\beta^*(X_{12}) = -X_{12}/\Delta$, $\beta^*(X_{21}) = -X_{21}/\Delta$, $\beta^*(X_{22}) = X_{11}/\Delta$, $\beta^*(1/\Delta) = \Delta$, we can readily see that β^* is a k-isomorphism.

10. The State Space: Realizations

We have, up to now, examined the relations between the transfer function, polynomial, and Hankel representations of a scalar linear system. We proved that these representations were "algebraically" the same. In this section, we shall consider the state space representation and examine its relation to the other representations. We begin by introducing the following:

Definition 10.1.   The triple (A, b, c) is a *(state space) realization* of the transfer function $f(z)$ if $f(z) = c(zI - A)^{-1}b$ or, equivalently, if $f(z) = \sum_{j=1}^{\infty}(cA^{j-1}b)z^{-j}$. Similarly, the triple (A, b, c) is a *(state space) realization* of the Hankel matrix $H = (h_{i+j-1})_{i,j=1}^{\infty}$ if $h_i = cA^{i-1}b$ for $i = 1, \ldots$.

We observe that if (A, b, c) is a realization of $f(z)$ (or of H) with $A \in M(n, n; k)$, then the degree of $f(z)$ (or of H) is at most n by the Cayley-Hamilton theorem. We shall view the triple (A, b, c) as a point in $A_k^{n^2+2n}$ and we consider a polynomial ring $k[A_k^{n^2+2n}]$ which we write $k[(X_{ij}), (Y_i), (Z_j)]$, $i, j = 1, \ldots, n$ or simply $k[X, Y, Z]$. We will define some morphisms of $A_k^{n^2+2n}$ into A_k^{2n} which we shall call *realization maps*.

First, let $B_0 \ldots, B_{n-1}, A_0, \ldots, A_{n-1}$ denote the coordinate functions on A_k^{2n} and define a polynomial mapping $\Re_f : A_k^{n^2+2n} \to A_k^{2n}$ by setting $\Re_f(X, Y, Z) = (B_0(X, Y, Z), \ldots, B_{n-1}(X, Y, Z), A_0(X, Y, Z), \ldots, A_{n-1}(X, Y, Z))$ where

$$f(z, X, Y, Z) = \frac{Z\operatorname{adj}(zI - X)Y}{\det(zI - X)} \tag{10.2}$$

$$= \frac{B_0(X, Y, Z) + \cdots + B_{n-1}(X, Y, Z)z^{n-1}}{A_0(X, Y, Z) + \cdots + A_{n-1}(X, Y, Z)z^{n-1} + z^n}$$

We call $\Re_f$ the *transfer function realization map*.

Example 10.3.   Let $n = 2$. Then

$$zI - X = \begin{bmatrix} z - X_{11} & -X_{12} \\ -X_{21} & z - X_{22} \end{bmatrix}$$

so that $\det(zI - X) = Z^2 - (X_{11} + X_{22})z + X_{11}X_{22} - X_{12}X_{21}$ and $A_0(X,Y,Z) = X_{11}X_{22} - X_{12}X_{21}$, $A_1(X,Y,Z) = -(X_{11} + X_{22})$. Also

$$\text{adj}(zI - X) = \begin{bmatrix} z - X_{22} & X_{12} \\ X_{21} & z - X_{11} \end{bmatrix}$$

so that $Z\,\text{adj}(zI-X)Y = \{(Z_1Y_1)+(Z_2Y_2)\}z+Z_1(-X_{22}Y_1+X_{12}Y_2)+Z_2(X_{21}Y_1 - X_{11}Y_2)$ and $B_0(X,Y,Z) = ZY \cdot (A_1(X,Y,Z)) + ZXY$, $B_1(X,Y,Z) = ZY$. Thus $\mathfrak{R}_f$ is a morphism. Let (b_0,b_1,a_0,a_1) be any element of $\mathbb{A}_k^4$. If we let

$$A = \begin{bmatrix} 0 & 1 \\ -a_0 & -a_1 \end{bmatrix} \;,\; b = \begin{bmatrix} 0 \\ 1 \end{bmatrix} \;,\; c = [\, b_0 \quad b_1 \,]$$

then $\mathfrak{R}_f(A,b,c) = (b_0,b_1,a_0,a_1)$ so that $\mathfrak{R}_f$ is surjective. If $g = (g_{ij})$ is an element of $GL(2,k)$, then $\det[zI-gXg^{-1}] = \det[g(zI-X)g^{-1}] = \det[zI - X]$ and $(Zg^{-1})(gY) = ZY$, $(Zg^{-1})(gXg^{-1})(gY) = ZXY$ so that $\mathfrak{R}_f(A,b,c) = \mathfrak{R}_f(gAg^{-1},gb,cg^{-1})$ for all (A,b,c). Thus, $\mathfrak{R}_f$ is not injective.

Now, let $H_1,\ldots,H_{2n}$ denote the coordinate functions on $\mathbb{A}_k^{2n}$ and define a polynomial mapping $\mathfrak{R}_h : \mathbb{A}_k^{n^2+2n} \to \mathbb{A}_k^{2n}$ by setting $\mathfrak{R}_h(X,Y,Z) = (H_1(X,Y,Z),\ldots,H_{2n}(X,Y,Z))$ where $H(X,Y,Z) = (ZX^{i+j-2}Y)_{i,j=1}^{\infty}$ so that

$$H_j(X,Y\,Z) = (ZX^{j-1}Y) \tag{10.4}$$

for $j = 1,\ldots,2n$. We call $\mathfrak{R}_h$ the *Hankel matrix realization map*.

Example 10.5.   Let $n = 2$. Then $H_j(X,Y,Z) = (ZX^{j-1}Y)$ for $j = 1,\ldots,4$ so that $H_1(X,Y,Z) = ZY = Z_1Y_1 + Z_2Y_2$, $H_2(X,Y,Z) = ZXY = Z_1X_{11}Y_1 + Z_1X_{12}Y_2 + Z_2X_{21}Y_1 + Z_2X_{22}Y_2$, etc. $\mathfrak{R}_h$ is clearly a morphism. We now observe that

$$X^2 = \begin{bmatrix} X_{11}^2 + X_{12}X_{21} & (X_{11} + X_{22})X_{12} \\ (X_{11} + X_{22})X_{21} & X_{12}X_{21} + X_{22}^2 \end{bmatrix}$$

so that

$$X^2 = (X_{11} + X_{22})X + (X_{12}X_{21} - X_{11}X_{22}) \cdot I$$

and hence that $H_3(X,Y,Z) = (ZX^2Y) = (X_{11} + X_{22})ZXY + (X_{12}X_{21}-X_{11}X_{22})ZY = (X_{11}+X_{22})H_2(X,Y,Z)+(X_{12}X_{21}-X_{11}X_{22})$

$H_1(X,Y,Z)$. Similarly, we have $H_4(X,Y,Z) = (X_{11}+X_{22})H_3(X,Y,Z)$ $+(X_{12}X_{21} - X_{11}X_{22})H_2(X,Y,Z)$. Let $(0,0,\xi,\eta)$ be an element of $\mathbb{A}_k^4$ with $\xi, \eta \neq 0$. Then $\mathfrak{R}_h^{-1}((0,0,\xi,\eta)) = \phi$ is empty i.e. $\mathfrak{R}_h$ is not surjective. We also note that $\mathfrak{R}_h(gAg^{-1},gb,cg^{-1}) = \mathfrak{R}_h(A,b,c)$ for all A,b,c where $g \in GL(2,k)$ so that $\mathfrak{R}_h$ is not injective.

Finally, let $(U_1,\ldots,U_n,\ldots,U_{2n})$ denote the coordinate functions on $\mathbb{A}_k^{2n}$ and define a polynomial mapping $\mathfrak{R}_\chi : \mathbb{A}_k^{n^2+2n} \to \mathbb{A}_k^{2n}$ by setting $\mathfrak{R}_\chi(X,Y,Z) = (U_1(X,Y,Z),\ldots,U_n(X,Y,Z),\ldots,U_{2n}(X,Y,Z))$ where

$$U_j(X,Y,Z) = ZX^{j-1}Y \qquad j = 1,\ldots,n$$
$$U_{n+t}(X,Y,Z) = -\chi_t(X) \qquad t = 1,\ldots,n \tag{10.6}$$

where $\det(zI - X) = z^n - \chi_1(X)z^{n-1} - \cdots - \chi_n(X)$. We call $\mathfrak{R}_\chi$ the *characteristic function realization map*.

Example 10.7.   Let $n = 2$. Then $U_1(X,Y,Z) = ZY$, $U_2(X,Y,Z) = ZXY$, $U_3(X,Y,Z) = -(X_{11} + X_{22})$ and $U_4(X,Y,Z) = X_{11}X_{22} - X_{21}X_{12}$. $\mathfrak{R}_\chi$ is clearly a morphism. Let $(\xi_1,\xi_2,\xi_3,\xi_4)$ be any element of $\mathbb{A}_k^4$ and let

$$A = \begin{bmatrix} 0 & 1 \\ -\xi_4 & -\xi_3 \end{bmatrix}, \quad b = \begin{bmatrix} 0 \\ 1 \end{bmatrix}, \quad c = [\xi_2 + \xi_1\xi_3 \quad \xi_1]$$

Then $\chi_1(A) = \xi_3$, $\chi_2(A) = \xi_4$, $U_1(A,b,c) = \xi_1$, and $U_2(A,b,c) = \xi_2$ so that $\mathfrak{R}_\chi$ is surjective. As before $\mathfrak{R}_\chi$ is not injective since $\mathfrak{R}_\chi(A,b,c) = \mathfrak{R}_\chi(gAg^{-1},gb,cg^{-1})$ for $g \in GL(2,k)$.

Proposition 10.8.   *The realization maps $\mathfrak{R}_f$, $\mathfrak{R}_h$ and $\mathfrak{R}_\chi$ are morphisms. Both $\mathfrak{R}_f$ and $\mathfrak{R}_\chi$ are surjective but $\mathfrak{R}_h$ is not. None of the maps is injective.*

Proof: (Left to reader.)

Now let us raise the following seemingly unrelated questions: (1) how are the realization maps related?, and (2) what is the structure of the open sets $\mathfrak{R}_f^{-1}(\mathrm{Rat}(n,k)) = \mathfrak{R}_f^{-1}((\mathbb{A}_k^{2n})_\rho)$ and $\mathfrak{R}_h^{-1}(\mathrm{Hank}(n,k)) = \mathfrak{R}_h^{-1}((\mathbb{A}_k^{2n})_\theta)$? To consider these questions, we introduce three maps ψ_f, ψ_h and ψ_χ of $\mathbb{A}_k^{2n}$ into itself. Let $(x_1,\ldots,x_{2n})$ be the coordinate functions on $\mathbb{A}_k^{2n}$. The map $\psi_f : \mathbb{A}_k^{2n} \to \mathbb{A}_k^{2n}$ is simply the Laurent map $\tilde{L}$ i.e.

$$\psi_f(x) = (\tilde{c}(x)\tilde{b},\ldots,\tilde{c}(x)\tilde{A}^{j-1}(x)\tilde{b},\ldots,\tilde{c}(x)\tilde{A}^{2n-1}(x)\tilde{b}) \tag{10.9}$$

where $\tilde{c}(x)$, $\tilde{A}(x)$ and $\tilde{b}$ are given by (8.5). Note that ψ_f is a morphism and that $\psi_f : (\mathbb{A}_k^{2n})_\rho \to (\mathbb{A}_k^{2n})_\theta$ is an isomorphism. The map $\psi_h : \mathbb{A}_k^{2n} \to \mathbb{A}_k^{2n}$ is given by $\psi_h(x_1, \ldots, x_{2n}) = (h_1(x), \ldots, h_{2n}(x))$ where

$$h_i(x) = x_i \qquad i = 1, \ldots, n$$

$$h_{n+j}(x) = -\sum_{i=1}^{n} x_{2n+1-i} h_{i+j-1}(x) \qquad j = 1, \ldots, n \qquad (10.10)$$

Note that ψ_h is a morphism. Finally, the map $\psi_\chi : \mathbb{A}_k^{2n} \to \mathbb{A}_k^{2n}$ is given by $\psi_\chi(x) = y$ where

$$y_r = \sum_{j=1}^{n-r} x_j x_{2n+1-j-r} + x_{n-r+1} \qquad r = 1, \ldots, n-1$$

$$y_n = x_1 \qquad (10.11)$$

$$y_{n+j} = x_{2n+1-j} \qquad j = 1, \ldots, n$$

Clearly ψ_χ is a morphism. The morphism ψ_f is neither surjective nor injective (since, say, for $n = 2$, $y_3 = -x_3 y_1 - x_4 y_2$, $y_4 = -x_3 y_2 - x_4 y_3$ and $\psi_f(0, 0, \xi, \eta) = (0, 0, 0, 0)$ for all ξ, η). Similarly, the morphism ψ_h is neither surjective nor injective. However, we do have:

Proposition 10.12. ψ_χ *is an isomorphism.*

Proof: Define a morphism of $\mathbb{A}_k^{2n} \to \mathbb{A}_k^{2n}$ by setting $x_1 = y_n, \ldots, x_r = y_{n-r+1} - \sum_{j=1}^{r-1} x_j(y) y_{2n+1+j-r}, r = 2, \ldots, n$, $x_{n+j} = y_{2n+1-j}, j = 1, \ldots, n$. A straightforward calculation shows that this morphism is ψ_χ^{-1}.

Example 10.13. Let $n = 3$. Then ψ_χ may be written

$$\begin{bmatrix} y_1 \\ y_2 \\ y_3 \\ y_4 \\ y_5 \\ y_6 \end{bmatrix} = \begin{bmatrix} x_5 & x_4 & 1 & 0 & 0 & 0 \\ x_4 & 1 & 0 & 0 & 0 & 0 \\ 1 & 0 & 0 & 0 & 0 & 0 \\ 0 & 0 & 0 & 0 & 0 & 1 \\ 0 & 0 & 0 & 0 & 1 & 0 \\ 0 & 0 & 0 & 1 & 0 & 0 \end{bmatrix} \begin{bmatrix} x_1 \\ x_2 \\ x_3 \\ x_4 \\ x_5 \\ x_6 \end{bmatrix}$$

If $\psi_\chi(\xi) = \psi_\chi(\xi')$, then $\xi_4 = y_6(\xi) = y_6(\xi') = \xi_4'$, $\xi_5 = y_5(\xi) = y_5(\xi') = \xi_5'$, $\xi_6 = y_4(\xi) = y_4(\xi') = \xi_6'$, $\xi_1 = y_3(\xi) = y_3(\xi') = \xi_1'$,

$\xi_2 = y_2(\xi) - \xi_4\xi_1 = y_2(\xi') - \xi_4'\xi_1' = \xi_2'$, and $\xi_3 = y_1(\xi) - \xi_5\xi_1 - \xi_4\xi_2 = y_1(\xi') - \xi_5'\xi_1' - \xi_4'\xi_2' = \xi_3'$. Thus, ψ_χ is injective. Also, if $\eta \in A_k^6$, then, setting $\xi_1 = \eta_3$, $\xi_2 = \eta_2 - \eta_3\eta_6$, $\xi_3 = \eta_1 - \eta_5\eta_3 - \eta_6(\eta_2 - \eta_3\eta_6)$, $\xi_4 = \eta_6$, $\xi_5 = \eta_5$, $\xi_6 = \eta_4$, we have $\psi_\chi(\xi) = \eta$ so that ψ_χ is surjective. The map ψ_χ^{-1} may be written

$$
\begin{bmatrix} x_1 \\ x_2 \\ x_3 \\ x_4 \\ x_5 \\ x_6 \end{bmatrix}
\begin{bmatrix} 0 & 0 & 1 & 0 & 0 & 0 \\ 0 & 1 & -y_6 & 0 & 0 & 0 \\ 1 & -y_6 & y_6^2 - y_5 & 0 & 0 & 0 \\ 0 & 0 & 0 & 0 & 0 & 1 \\ 0 & 0 & 0 & 0 & 1 & 0 \\ 0 & 0 & 0 & 1 & 0 & 0 \end{bmatrix}
\begin{bmatrix} y_1 \\ y_2 \\ y_3 \\ y_4 \\ y_5 \\ y_6 \end{bmatrix}
$$

For instance, $x_2 = y_2 - y_3y_6 = x_4x_1 + x_2 - x_1x_4$ and $x_3 = y_1 - y_6y_2 + (y_6^2 - y_5)y_3 = x_3 + x_2x_4 + x_1x_5 - x_4(x_4x_1 + x_2) + (x_4^2 - x_5)x_1 = x_3$ so that $\psi_\chi^{-1} \circ \psi_\chi = I$.

Proposition 10.14.   *The realization maps $\mathfrak{R}_f$, $\mathfrak{R}_h$, $\mathfrak{R}_\chi$ and the associated morphisms ψ_f, ψ_h, ψ_χ satisfy the following relations:*

$$\psi_\chi \circ \mathfrak{R}_\chi = \mathfrak{R}_f$$
$$\psi_f \circ \mathfrak{R}_f = \mathfrak{R}_h$$
$$\psi_h \circ \mathfrak{R}_\chi = \mathfrak{R}_h$$
$$\psi_h = \psi_f \circ \psi_\chi$$

Proof: We first show that $\psi_\chi \circ \mathfrak{R}_\chi = \mathfrak{R}_f$. Let $(\psi_\chi \circ \mathfrak{R}_\chi)(A, b, c) = (y_1, \ldots, y_{2n})$. Then $y_r = -\sum_{j=1}^{n-r}(cA^{j-1}b)\chi_{n+1-(j+r)} + cA^{n-r}b$, for $r = 1, \ldots, n-1$, $y_n = cb$, and $y_{n+j} = -\chi_{n-j+1}$ for $j = 1, \ldots, n$. If $\mathfrak{R}_f(A, b, c) = (b_0, \ldots, b_{n-1}, a_0, \ldots, a_{n-1})$, then $a_j = -\chi_{n-j}$, $j = 0, \ldots, n-1$, $b_{n-1} = cb$, and $b_r = \sum_{j=1}^{n-r-1}(cA^{j-1}b)a_{j+r} + cA^{n-r-1}b$, $r = n-2, n-1, \ldots, 0$. It follows that $a_j = y_{n+1+j}$, $j = 0, \ldots, n-1$, $b_{n-1} = y_n$, and $b_r = y_{r+1}$, $r = 0, \ldots, n-2$. In other words, $\psi_\chi \circ \mathfrak{R}_\chi = \mathfrak{R}_f$. To show that $\psi_h \circ \mathfrak{R}_\chi = \mathfrak{R}_h$, we observe that, by the Cayley-Hamilton Theorem, $A^{n+j-1} = \sum_{i=1}^n \chi_{n+1-i}A^{i+j-2}$, $j = 1, \ldots, n$, and hence, that $\psi_h(\mathfrak{R}_\chi(A, b, c)) = (cb, \ldots, cA^{n-1}b, \ldots, \sum_{i=1}^n \chi_{n+1-i}h_{i+j-1}(x), \ldots) = (cb, \ldots, cA^{n-1}b, \ldots cA^{j-1}b, \ldots, cA^{2n-1}b) = \mathfrak{R}_h(A, b, c)$. If we show that $\psi_h = \psi_f \circ \psi_\chi$, then $\psi_f \circ \mathfrak{R}_f = \psi_f \circ \psi_\chi \circ \mathfrak{R}_\chi = \psi_h \circ \mathfrak{R}_\chi = \mathfrak{R}_h$.

In view of the fact that $\chi_i(\tilde{A}(y)) = -y_{2n+1-i}$ (where $\tilde{A}(y)$ is given by (8.5)), we have $\chi_i(\tilde{A}(\psi_\chi(x))) = -x_{n+i}$, $i = 1, \ldots, n$, and hence, that

$$\tilde{c}(\psi_\chi(x))\tilde{A}^{n+j-1}(\psi_\chi(x))\tilde{b} = \sum_{i=1}^{n} \chi_{n+1-i}(\tilde{A}(\psi_\chi(x))) \cdot$$

$$\tilde{c}(\psi_\chi(x))\tilde{A}^{i+j-2}(\psi_\chi(x))\tilde{b} = -\sum_{i=1}^{n} x_{2n+1-i}\tilde{c}(\psi_\chi(x))\tilde{A}^{i+j-2}(\psi_\chi(x))\tilde{b}$$

for $j = 1, \ldots, n$. In other words, to show that $\psi_h = \psi_f \circ \psi_\chi$ it will be enough to show that $\tilde{c}(\psi_\chi(x))\tilde{A}^{j-1}(\psi_\chi(x))\tilde{b} = x_j$ for $j = 1, \ldots, n$. But $\tilde{c}(\psi_\chi(x)) = [x_1 \ldots x_n]g_u$ and $\tilde{A}(\psi_\chi(x)) = A_u$ where $u_1 = x_{2n}, \ldots, u_n = x_{n+1}$ and g_u, A_u are of the form given in lemma 8.13. The result then follows from lemma 8.18.

Corollary 10.15. *Let* $\lambda : \mathbb{A}_k^{2n} \to k$ *be given by* $\lambda = \psi_\chi^*(\rho) = \rho \circ \psi_\chi$. *Then* $\psi_\chi : (\mathbb{A}_k^{2n})_\lambda \to (\mathbb{A}_k^{2n})_\rho$ *and* $\psi_h : (\mathbb{A}_k^{2n})_\lambda \to (\mathbb{A}_k^{2n})_\theta$ *are k-isomorphisms. Thus, the diagram*

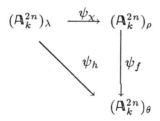

is a commutative diagram of k-isomorphisms.

Corollary 10.16. *The principal affine open sets* $\mathfrak{R}_f^{-1}((\mathbb{A}_k^{2n})_\rho)$, $\mathfrak{R}_h^{-1}((\mathbb{A}_k^{2n})_\theta)$ *and* $\mathfrak{R}_\chi^{-1}((\mathbb{A}_k^{2n})_\lambda)$ *are all the same.*

Proof: $\mathfrak{R}_f^{-1}((\mathbb{A}_k^{2n})_\rho) = \{\xi : (\rho \circ \mathfrak{R}_f)(\xi) \neq 0\}$, $\mathfrak{R}_h^{-1}((\mathbb{A}_k^{2n})_\theta) = \{\xi : (\theta \circ \mathfrak{R}_h)(\xi) \neq 0\}$, and $\mathfrak{R}_\chi^{-1}((\mathbb{A}_k^{2n})_\lambda) = \{\xi : (\lambda \circ \mathfrak{R}_\chi)(\xi) \neq 0\}$. But $\mathfrak{R}_f = \psi_\chi \circ \mathfrak{R}_\chi$ and so, $\rho \circ \mathfrak{R}_f = \rho \circ \psi_\chi \circ \mathfrak{R}_\chi = \lambda \circ \mathfrak{R}_\chi$ as $\lambda = \rho \circ \psi_\chi$. Similarly, $\mathfrak{R}_h = \psi_h \circ \mathfrak{R}_\chi$ and $\theta \circ \mathfrak{R}_h = (\theta \circ \psi_h) \circ \mathfrak{R}_\chi$ with $\theta \circ \psi_h = (\theta \circ \psi_f) \circ \psi_\chi$, $-(\rho \circ \psi_\chi) = -\lambda$ since $(\theta \circ \psi_f) = -\rho$.

Example 10.17. Let $n = 2$. Then the maps ψ_f, ψ_h, ψ_χ are given by:

$$\psi_f(x_1, x_2, x_3, x_4) = (x_2, x_1 - x_4 x_2, -x_3 x_2 - x_4(x_1 - x_4 x_2), \alpha(x)),$$
$$\alpha(x) = -x_3(x_1 - x_4 x_2) - x_4(-x_3 x_2 - x_4(x_1 - x_4 x_2)))$$
$$\psi_h(x_1, x_2, x_3, x_4) = (x_1, x_2, -x_4 x_1 - x_3 x_2, -x_4 x_2 - x_3(-x_4 x_1 - x_3 x_2))$$
$$\psi_\chi(x_1, x_2, x_3, x_4) = (x_1 x_3 + x_2, x_1, x_4, x_3)$$

The functions ρ, θ, λ are given by:

$$\rho(x_1, x_2, x_3, x_4) = x_1^2 - x_4 x_1 x_2 + x_3 x_2^2$$
$$\theta(x_1, x_2, x_3, x_4) = x_1 x_3 - x_2^2$$
$$\lambda(x_1, x_2, x_3, x_4) = x_2^2 + x_1 x_2 x_3 + x_4 x_1^2$$

Clearly $\rho \circ \psi_\chi = \lambda$. Now $(\theta \circ \psi_h)(x_1, x_2, x_3, x_4) = \theta(x_1, x_2, -x_4 x_1 - x_3 x_2, -x_4 x_2 - x_3(-x_4 x_1 - x_3 x_2)) = x_1(-x_4 x_1 - x_3 x_2) - x_2^2 = -x_4 x_1^2 - x_1 x_2 x_3 - x_2^2 = \lambda(x_1, x_2, x_3, x_4)$. Thus we can see that ψ_h maps $(\mathbb{A}_k^{2n})_\lambda$ into $(\mathbb{A}_k^{2n})_\theta$. Let us show directly that ψ_h is bijective as a map of $(\mathbb{A}_k^{2n})_\lambda$ into $(\mathbb{A}_k^{2n})_\theta$. If $\psi_h(\xi_1, \xi_2, \xi_3, \xi_4) = \psi_h(\xi_1', \xi_2', \xi_3', \xi_4')$, then $\xi_1 = \xi_1'$, $\xi_2 = \xi_2'$, $-\xi_4 \xi_1 - \xi_3 \xi_2 = -\xi_4' \xi_1' - \xi_3' \xi_2'$, and $-\xi_4 \xi_2 - \xi_3(-\xi_4 \xi_1 - \xi_3 \xi_2) = -\xi_4' \xi_2' - \xi_3'(-\xi_4' \xi_1' - \xi_3' \xi_2')$. It follows that

$$(\xi_4' - \xi_4)\xi_1 - (\xi_3' - \xi_3)\xi_2 = 0$$
$$(\xi_4' - \xi_4)\xi_2 - (\xi_3' - \xi_3)(-\xi_4 \xi_1 - \xi_3 \xi_2) = 0$$

and, hence, that $\xi_4' = \xi_4$, $\xi_3' = \xi_3$ since $\lambda(\xi_1, \xi_2, \xi_3, \xi_4) = \xi_2^2 + \xi_1 \xi_2 \xi_3 + \xi_1^2 \xi_4 \neq 0$. In other words, ψ_h is injective. If $(\eta_1, \eta_2, \eta_3, \eta_4) \in (\mathbb{A}_k^{2n})_\theta$ and if we let $\xi_1 = \eta_1$, $\xi_2 = \eta_2$, $\xi_3 = (\eta_2 \eta_3 - \eta_1 \eta_4)/(\eta_1 \eta_3 - \eta_2^2)$, $\xi_4 = (\eta_2 \eta_4 - \eta_3^2)/(\eta_1 \eta_3 - \eta_2^2)$, then $\psi_h(\xi_1, \xi_2, \xi_3, \xi_4) = (\eta_1, \eta_2, (-\eta_1 \eta_2 \eta_4 + \eta_1 \eta_3^2 - \eta_2^2 \eta_3 + \eta_1 \eta_2 \eta_4)/(\eta_1 \eta_3 - \eta_2^2), (-\eta_2^2 \eta_4 + \eta_2 \eta_3^2 - \eta_2 \eta_3^2 + \eta_1 \eta_3 \eta_4)/(\eta_1 \eta_3 - \eta_2^2)) = (\eta_1, \eta_2, \eta_3, \eta_4)$ and ψ_h is surjective. Since $\psi_h^{-1} = \psi_\chi^{-1} \circ \psi_f^{-1}$, ψ_h is a k-isomorphism.

We let $S_{1,1}^n$ be the principal affine open set $\mathfrak{R}_\chi^{-1}(\mathbb{A}_k^{2n})_\lambda = \mathfrak{R}_f^{-1}(\mathbb{A}_k^{2n})_\rho = \mathfrak{R}_h^{-1}(\mathbb{A}_k^{2n})_\theta$ and we call $S_{1,1}^n$ the *set of linear systems of degree* n. We should like to characterize the elements of $S_{1,1}^n$. Denoting $(\mathbb{A}_k^{2n})_\lambda$ by $\text{Char}(n, k)$, we observe that if (A, b, c) is an element of $S_{1,1}^n$, then $\mathfrak{R}_f(A, b, c)$ is an element of $\text{Rat}(n, k)$, $\mathfrak{R}_h(A, b, c)$ is an element of $\text{Hank}(n, k)$, and $\mathfrak{R}_\chi(A, b, c)$ is an element of $\text{Char}(n, k)$. Since $\mathfrak{R}_f$ and $\mathfrak{R}_\chi$ are surjective and ψ_f is an isomorphism between $(\mathbb{A}_k^{2n})_\rho$ and $(\mathbb{A}_k^{2n})_\theta$, we can see that, conversely, given an element $f(z)$ of $\text{Rat}(n,k)$ or H of $\text{Hank}(n,k)$ or ξ of $\text{Char}(n,k)$, there is an element (A,b,c) of $S_{1,1}^n$ such that $\mathfrak{R}_f(A,b,c)\text{``} = \text{''}f(z)$ or $\mathfrak{R}_h(A,b,c)\text{``} = \text{''}H$ or $\mathfrak{R}_\chi(A,b,c)\text{``} = \text{''}\xi$. In other words, if $f(z)$ is a transfer function of degree n or if H is a Hankel matrix of degree n or if ξ is a characteristic representation of degree n, then there is a linear system of degree n which is a *realization of $f(z)$ or of H or of ξ*. We summarize this in the following:

Theorem 10.18. *If $x = (A,b,c)$ is an element of $S_{1,1}^n$, then $f_x(z) = c(zI - A)^{-1}b$ is an element of $\text{Rat}(n,k)$. Conversely, if $f(z)$ is an element of $\text{Rat}(n,k)$, then there is an $x = (A,b,c)$ in $S_{1,1}^n$ such that $f_x(z) = f(z)$. Similarly, if $x = (A,b,c)$ is an element of $S_{1,1}^n$, then $H_x = (cA^{i+j-2}b)_{i,j=1}^\infty$ is an element of $\text{Hank}(n,k)$ and conversely, if H is an element of $\text{Hank}(n,k)$, then there is an $x = (A,b,c)$ in $S_{1,1}^n$ with $H_x = H$. Finally, if $x = (A,b,c)$ is an element of $S_{1,1}^n$, then $\xi_x = (cb,\ldots,cA^{n-1}b,-\chi_1(A),\ldots,-\chi_n(A))$ is an element of $\text{Char}(n,k)$ and conversely, if ξ is an element of $\text{Char}(n,k)$, then there is an $x = (A,b,c)$ in $S_{1,1}^n$ with $\xi_x = \xi$. Such an x is called a minimal realization of $f(z)$ or of H or of ξ.*

We have thus established that the set of linear systems of degree n, $S_{1,1}^n$, is the set of minimal realizations for $\text{Rat}(n,k)$ or $\text{Hank}(n,k)$ or $\text{Char}(n,k)$. In other words, $\mathfrak{R}_f : S_{1,1}^n \to \text{Rat}(n,k)$, $\mathfrak{R}_h : S_{1,1}^n \to \text{Hank}(n,k)$, and $\mathfrak{R}_\chi : S_{1,1}^n \to \text{Char}(n,k)$ are *surjective* morphisms.

11. The State Space:
Controllability, Observability, Equivalence

We noted in section 10 that the various realization maps were never injective. We should like to examine this issue and to further characterize the set of linear systems of degree n.

We again view the triple (A, b, c) as an element of $\mathbb{A}_k^{n^2+2n}$ and we consider the polynomial ring $k[\mathbb{A}_k^{n^2+2n}] = k[(X_{ij}), (Y_i), (Z_j)] = k[X, Y, Z]$. We will define some morphisms of $\mathbb{A}_k^{n^2+2n}$ into $\mathbb{A}_k^{n^2}$ which we shall call the *controllability* and *observability* maps.

Let $\mathbf{Y} : \mathbb{A}_k^{n^2+2n} \to \mathbb{A}_k^{n^2}$ be the map given by

$$\mathbf{Y}(X, Y, Z) = [Y \ XY \cdots X^{n-1}Y] \tag{11.1}$$

The morphism $\mathbf{Y}$ is called the *controllability map*. Similarly, let $\mathbf{Z} : \mathbb{A}_k^{n^2+2n} \to \mathbb{A}_k^{n^2}$ be the map given by

$$\mathbf{Z}(X, Y, Z) = \begin{bmatrix} Z \\ ZX \\ \vdots \\ ZX^{n-1} \end{bmatrix} \tag{11.2}$$

The morphism $\mathbf{Z}$ is called the *observability map*. We note that $\mathbf{Y}$ is independent of Z and that $\mathbf{Z}$ is independent of Y. Finally, we let $\mathbf{H} : \mathbb{A}_k^{n^2+2n} \to \mathbb{A}_k^{2n}$ be given by

$$\mathbf{H}(X, Y, Z) = \mathbf{Z}(X, Y, Z) \cdot \mathbf{Y}(X, Y, Z) \tag{11.3}$$

and we call the morphism $\mathbf{H}$ the *Hankel matrix map*.

Example 11.4.   Let $n = 2$. Then

$$\mathbf{Y}(X,Y,Z) = \begin{bmatrix} Y_1 & X_{11}Y_1 + X_{12}Y_2 \\ Y_2 & X_{21}Y_1 + X_{22}Y_2 \end{bmatrix}$$

$$\mathbf{Z}(X,Y,Z) = \begin{bmatrix} Z_1 & Z_2 \\ Z_1X_{11} + Z_2X_{21} & Z_1X_{12} + Z_{22}X_{22} \end{bmatrix}$$

and

$$\mathbf{H}(X,Y,Z) = \begin{bmatrix} ZY & ZXY \\ ZXY & ZX^2Y \end{bmatrix}$$

We note that rank $[\mathbf{H}(A,b,c)] \le \min[\text{rank } \mathbf{Y}(A,b,c), \text{rank } \mathbf{Z}(A,b,c)] \le n$ for all (A,b,c) in $\mathbb{A}_k^{n^2+2n}$. This leads us to make the following definition:

Definition 11.5.   The scalar linear system (A,b,c) is *controllable* if rank $\mathbf{Y}(A,b,c) = n$ and *observable* if rank $\mathbf{Z}(A,b,c) = n$. If rank $\mathbf{H}(A,b,c) = n$, then (A,b,c) is *minimal*.

We note that a system is minimal if and only if it is both controllable and observable. Letting $\gamma(X,Y,Z) = \det \mathbf{Y}(X,Y,Z)$ and $\omega(X,Y,Z) = \det \mathbf{Z}(X,Y,Z)$, we can see that γ and ω are elements of $k[X,Y,Z]$. If $I_c = (\gamma)$ and $I_o = (\omega)$, then $V_c = V(I_c)$ and $V_o = V(I_o)$ are affine algebraic sets and $V_c \cup V_o = V(I_cI_o) = V(I_c \cap I_o) = V(I_\theta) = V_h$ where $\theta(X,Y,Z) = \omega(X,Y,Z) \cdot \gamma(X,Y,Z) = \det \mathbf{H}(X,Y,Z)$. It follows immediately that $x = (A,b,c)$ is controllable if and only if $x \notin V_c$; that $x = (A,b,c)$ is observable if and only if $x \notin V_o$; and, that $x = (A,b,c)$ is minimal if and only if $x \notin V_c \cup V_o = V_h$. In other words, the controllable systems are the elements of the principal affine open set $(\mathbb{A}_k^{n^2+2n})_\gamma$; the observable systems are the elements of the principal affine open set $(\mathbb{A}_k^{n^2+2n})_\omega$; and, the minimal systems are the elements of the principal affine open set $(\mathbb{A}_k^{n^2+2n})_\theta$.

Proposition 11.6. $(\mathbb{A}_k^{n^2+2n})_\theta = S_{1,1}^n$ *(so that both notions of a minimal linear system are the same)*.

Proof: Simply observe that $(\theta \circ \mathfrak{R}_h)(X,Y,Z) = \theta(X,Y,Z)$. Thus, if $(A,b,c) \in S_{1,1}^n$, then $\mathfrak{R}_h(A,b,c) \in (\mathbb{A}_k^{2n})_\theta$ and $(\theta \circ \mathfrak{R}_h)(A,b,c) =$

$\theta(A, b, c) \neq 0$. Conversely, if $\theta(A, b, c) \neq 0$, then $\mathfrak{R}_h(A, b, c) \in (\mathbb{A}_k^{2n})_\theta$ and $(A, b, c) \in S_{1,1}^n$.

Thus, we may characterize $S_{1,1}^n$ as the set of systems which are both controllable and observable.

Now, let us ask the following question: when do two minimal triples $x = (A, b, c)$ and $x_1 = (A_1, b_1, c_1)$ generate the same $f(z)$ in $\mathrm{Rat}(n, k)$ or H in $\mathrm{Hank}(n, k)$ or ξ in $\mathrm{Char}(n, k)$? In order to answer this question, we shall introduce an equivalence in $\mathbb{A}_k^{n^2+2n}$ based on the "action" of a group of transformations.

Definition 11.7. Let $GL(n, k) = \{g : g \in M(n, n; k), \det g \neq 0\}$. $GL(n, k)$ is called the *general linear group over* k.

We observe that since $M(n, n; k)$ may be identified with $\mathbb{A}_k^{n^2}$, $GL(n, k)$ may be identified with the principal affine open subset defined by the nonvanishing of the determinant which is a polynomial in the entries of a matrix.

Proposition 11.8. *Let $x = (A, b, c)$ be an element of $\mathbb{A}_k^{n^2+2n}$ and let g be an element of $GL(n, k)$. If $x_1 = (A_1, b_1, c_1)$ where*

$$A_1 = gAg^{-1}, \quad b_1 = gb, \quad c_1 = cg^{-1} \qquad (11.9)$$

then x_1 is an element of $\mathbb{A}_k^{n^2+2n}$ (obviously) and

$$\mathfrak{R}_f(x) = \mathfrak{R}_f(x_1), \quad \mathfrak{R}_h(x) = \mathfrak{R}_h(x_1), \quad \mathfrak{R}_\chi(x) = \mathfrak{R}_\chi(x_1) \qquad (11.10)$$

for the realization maps $\mathfrak{R}_f$, $\mathfrak{R}_h$ and $\mathfrak{R}_\chi$.

Proof: To show that $\mathfrak{R}_f(x) = \mathfrak{R}_f(x_1)$, we simply note that $f_x(z) = c(zI - A)^{-1}b = (cg^{-1})g[zg^{-1}g - g^{-1}gAg^{-1}g]^{-1}g^{-1}(gb) = c_1(zI - A_1)^{-1}b_1 = f_{x_1}(z)$. As for showing that $\mathfrak{R}_h(x) = \mathfrak{R}_h(x_1)$, we have

$$c_1 A_1^{j-1} b_1 = (cg^{-1})(gAg^{-1})^{j-1}(gb)$$
$$= (cg^{-1})gA^{j-1}g^{-1}(gb) = cA^{j-1}b$$

for $j = 1, 2, \ldots$. Finally, in view of what was just proved, we need only observe that $\chi_i(A_1) = \chi_i(gAg^{-1}) = \chi_i(A)$, $i = 1, \ldots, n$, in order to establish that $\mathfrak{R}_\chi(x) = \mathfrak{R}_\chi(x_1)$.

We say that $GL(n,k)$ "acts" on $A_k^{n^2+2n}$ via (11.9) and we write $g \cdot x = g \cdot (A,b,c)$ for this "action" so that

$$g \cdot x = g \cdot (A,b,c) = (gAg^{-1}, gb, cg^{-1}) \qquad (11.11)$$

In effect, the action of $GL(n,k)$ corresponds to coordinate change in the state space. Now, noting that $GL(n,k)$ is a quasi-affine variety and that $A_k^{n^2+2n}$ is an affine variety, we may view the action (11.11) as a mapping ψ_G of $GL(n,k) \times A_k^{n^2+2n}$ into $A_k^{n^2+2n}$ i.e., letting $G = GL(n,k)$, we have $\psi_G : G \times A_k^{n^2+2n} \to A_k^{n^2+2n}$ given by

$$\psi_G(g,x) = g \cdot x = (gAg^{-1}, gb, cg^{-1}) \qquad (11.12)$$

where $x = (A,b,c)$. Now, if $G \times A_k^{n^2+2n}$ were an affine variety i.e. if there were a notion of product, then we could ask whether or not ψ_G was a morphism and what were the algebraic properties of ψ_G. For this and many other reasons, it is necessary to have a notion of "product". We shall postpone discussion of this concept until section 12, and now turn our attention to examining the effect of the action of $GL(n,k) = G$ on the controllable, observable and minimal linear systems.

Proposition 11.13. *If $x = (A,b,c)$ is an element of $A_k^{n^2+2n}$ and g is an element of $G = GL(n,k)$, then*

$$\begin{aligned} \mathbf{Y}(g \cdot x) &= g\mathbf{Y}(x) \\ \mathbf{Z}(g \cdot x) &= \mathbf{Z}(x)g^{-1} \end{aligned} \qquad (11.14)$$

where $\mathbf{Y}(x) = \mathbf{Y}(A,b,c)$, $\mathbf{Z}(x) = \mathbf{Z}(A,b,c)$ and $\mathbf{Y}, \mathbf{Z}$ are the controllability and observability maps respectively.

Proof: Simply note that

$$\begin{aligned} \mathbf{Y}(g \cdot x) &= [gb \ gAg^{-1}(gb) \cdots (gAg^{-1})^{n-1}(gb)] \\ &= [gb \ gAb \cdots gA^{n-1}b] = g[b \ Ab \cdots A^{n-1}b] \\ &= g\mathbf{Y}(x) \end{aligned}$$

and similarly, for $\mathbf{Z}(g \cdot x)$.

Corollary 11.15. *If* $x = (A, b, c)$ *is controllable, or observable, or minimal and* g *is an element of* G, *then* $g \cdot x$ *is controllable, or observable, or minimal.*

Corollary 11.16. *If* $x = (A, b, c)$ *is controllable (or observable) and* g *is an element of* G *such that* $g \cdot x = x$, *then* $g = I$ *is the identity.*

Proof: Since $\mathbf{Y}(x) = \mathbf{Y}(g \cdot x) = g\mathbf{Y}(x)$ and $\mathbf{Y}(x)$ is invertible by controllability, we have $g = I$.

Example 11.17. Let $n = 2$ and let

$$A = \begin{bmatrix} 1 & 0 \\ 0 & 1 \end{bmatrix} \; , \; b = \begin{bmatrix} 0 \\ 1 \end{bmatrix} \; , \; c = [1 \quad 0]$$

so that (A, b, c) is neither controllable nor observable. If t is *any* element of k and if $g_t = \begin{bmatrix} 1 & 0 \\ t & 1 \end{bmatrix}$ so that $g_t^{-1} = \begin{bmatrix} 1 & 0 \\ -t & 1 \end{bmatrix}$, then clearly $g_t \cdot x = x$.

Corollary 11.18. *If* $x = (A, b, c)$ *is controllable (or observable) and* $g, \tilde{g}$ *are elements of* G *such that* $g \cdot x = \tilde{g} \cdot x$, *then* $g = \tilde{g}$.

Proof: If $g \cdot x = \tilde{g} \cdot x$, then $gAg^{-1} = \tilde{g}A\tilde{g}^{-1}$, $gb = \tilde{g}b$, and $cg^{-1} = c\tilde{g}^{-1}$ so that $(\tilde{g}^{-1}g)A(\tilde{g}^{-1}g)^{-1} = A$, $(\tilde{g}^{-1}g)b = b$ and $c(\tilde{g}^{-1}g)^{-1} = c$. In other words, $(\tilde{g}^{-1}g) \cdot x = x$ and so, $\tilde{g}^{-1}g = I$ or $g = \tilde{g}$.

We are now ready to state and prove a second major theorem, namely the State Space Isomorphism theorem. This theorem essentially states that minimal realizations of the same transfer function or Hankel matrix or element of $\text{Char}(n, k)$ are "equivalent" under the action of $GL(n, k)$ with a "unique equivalence". More precisely, we have:

Theorem 11.19.   (State Space Isomorphism Theorem)

(1) *Let* $f(z)$ *be an element of* $\text{Rat}(n, k)$ *and let* $x = (A, b, c)$ *and* $x_1 = (A_1, b_1, c_1)$ *be minimal realizations of* $f(z)$. *Then there is a* unique $g \in GL(n, k)$ *such that* $g \cdot x = x_1$.

(2) *Let* H *be an element of* $\text{Hank}(n, k)$ *and let* $x = (A, b, c)$ *and let* $x_1 = (A_1, b_1, c_1)$ *be minimal realizations of* H. *Then there is a* unique $g \in GL(n, k)$ *such that* $g \cdot x = x_1$.

(3) *Let ξ be an element of* $\mathrm{Char}(n,k)$ *and let* $x = (A,b,c)$ *and* $x_1 = (A_1,b_1,c_1)$ *be minimal realizations of* ξ. *Then there is a* unique $g \in GL(n,k)$ *such that* $g \cdot x = x_1$.

Proof: We first note that, in all three cases, if $g \cdot x = x_1$ and $\tilde{g} \cdot x = x_1$ for $g, \tilde{g} \in G = GL(n,k)$, then $g = \tilde{g}$ by corollary 11.18. In other words, uniqueness is established.

Next, we note that, in view of proposition 10.14 and proposition 10.12, $\Re_\chi(x) = \Re_\chi(x_1)$ if and only if

$$\Re_f(x) = (\psi_\chi \circ \Re_\chi)(x) = (\psi_\chi \circ \Re_\chi)(x_1) = \Re_f(x_1).$$

Similarly, if $\Re_f(x) = \Re_f(x_1)$, then $\Re_h(x) = (\psi_f \circ \Re_f)(x) = (\psi_f \circ \Re_f)(x_1) = \Re_h(x_1)$. If, on the other hand, $\Re_h(x) = \Re_h(x_1)$ and x, x_1 are elements of $S_{1,1}^n$, then

$$\Re_f(x) = (\psi_f^{-1} \circ \Re_h)(x) = (\psi_f^{-1} \circ \Re_h)(x_1) = \Re_f(x_1)$$

in view of corollary 10.15. In other words, it will be enough to establish existence for the case (2).

So suppose that H is an element of $\mathrm{Hank}(n,k)$ and let $x = (A,b,c)$ and $x_1 = (A_1,b_1,c_1)$ be minimal realizations of H. Then $\Re_h(x) = \Re_h(x_1)$ so that

$$cA^{j-1}b = c_1 A_1^{j-1} b_1 \tag{11.20}$$

for $j = 1, \ldots,$ and

$$\mathbf{Z}(A,b,c)\mathbf{Y}(A,b,c) = \mathbf{H}(A,b,c) \tag{11.21}$$
$$= \mathbf{H}(A_1,b_1,c_1) = \mathbf{Z}(A_1,b_1,c_1)\mathbf{Y}(A_1,b_1,c_1)$$

Let $\mathbf{Z} = \mathbf{Z}(A,b,c)$, $\mathbf{Y} = \mathbf{Y}(A,b,c)$, $\mathbf{Z}_1 = \mathbf{Z}(A_1,b_1,c_1)$, and $\mathbf{Y}_1 = \mathbf{Y}(A_1,b_1,c_1)$. By minimality, $\mathbf{Z}, \mathbf{Y}, \mathbf{Z}_1$ and $\mathbf{Y}_1$ are all nonsingular. But (11.20) and (11.21) imply that

$$\mathbf{Z}A\mathbf{Y} = \mathbf{Z}_1 A_1 \mathbf{Y}_1, \quad \mathbf{Z}b = \mathbf{Z}_1 b_1, \quad c\mathbf{Y} = c_1 \mathbf{Y}_1 \tag{11.22}$$

It follows from (11.21) that

$$\mathbf{Z}_1^{-1}\mathbf{Z} = \mathbf{Y}_1\mathbf{Y}^{-1} \tag{11.23}$$

and so, letting $g = Z_1^{-1}Z$, we have $gAg^{-1} = A_1$, $gb = b_1$, and $cg^{-1} = c_1$ in view of (11.22). In other words, $g \cdot x = x_1$.

Example 11.24.   Let $n = 2$ and let

$$A = \begin{bmatrix} 1 & 0 \\ 0 & 1 \end{bmatrix}, \quad b = \begin{bmatrix} 0 \\ 1 \end{bmatrix}, \quad c = [0 \quad 0]$$

$$A_1 = \begin{bmatrix} 1 & 1 \\ 0 & 1 \end{bmatrix}, \quad b_1 = \begin{bmatrix} 0 \\ 1 \end{bmatrix}, \quad c_1 = [0 \quad 0]$$

and $x = (A, b, c)$, $x_1 = (A_1, b_1, c_1)$. Then $\Re_\chi(x) = (0, 0, -2, 1) = \Re_\chi(x_1)$ and $\Re_h(x) = (0, 0, 0, 0) = \Re_h(x_1)$. However, there does not exist a g in $GL(2, k)$ with $g \cdot x = x_1$.

We have shown that elements x, x_1 of $S_{1,1}^n$ correspond to the same element $f(z)$ of $\mathrm{Rat}(n, k)$ (or H of $\mathrm{Hank}(n, k)$ or ξ of $\mathrm{Char}(n, k)$) if and only if there is a g in $GL(n, k)$ such that $g \cdot x = x_1$. In particular, we have observed that the group $G = GL(n, k)$ acts on $A_k^{n^2 + 2n}$ via

$$g \cdot (A, b, c) = (gAg^{-1}, gb, cg^{-1}) \tag{11.25}$$

and that $G \cdot S_{1,1}^n = S_{1,1}^n$ so that $S_{1,1}^n$ is "invariant" under that action of G. If we say that x, x_1 in $S_{1,1}^n$ are *equivalent under G* (or *modulo G*) when there is a g in G with $g \cdot x = x_1$, then the set of equivalence classes, denoted by $S_{1,1}^n/G$, can be identified with $\mathrm{Hank}(n, k)$ (or $\mathrm{Rat}(n, k)$ or $\mathrm{Char}(n, k)$). We shall eventually show that this identification can be made algebraic. In order to do so, we again require some algebra.

12. Affine Algebraic Geometry: Products, Graphs and Projections

We wish to define the product $V \times W$ of any two affine algebraic sets. We first observe that if $\xi = (\xi_1, \ldots, \xi_N) \in A_k^N$ and $\eta = (\eta_1, \ldots, \eta_M) \in A_k^M$, then $(\xi, \eta) = (\xi_1, \ldots, \xi_N, \eta_1, \ldots, \eta_M) \in A_k^{N+M}$ and A_k^{N+M} can be identified with the set of pairs (ξ, η) i.e. $A_k^{N+M} = A_k^N \times A_k^M$ as *sets*. However, the Zariski topology in A_k^{N+M} is not the same as the product of the Zariski topologies in A_k^N and A_k^M.

Example 12.1. Let $N = M = 1$. Then the set $\xi^2 - \eta = 0$ is closed in A_k^2 but is not closed in the product of the topologies in A_k^1 as it is not a finite union of lines parallel to the axes.

We note that we may identify $k[A_k^N]$ as a subring $k[X_1, \ldots, X_N]$ of $k[A_k^{N+M}] = k[X_1, \ldots, X_N, Y_1, \ldots, Y_M]$ and $k[A_k^M]$ with the subring $k[Y_1, \ldots, Y_M]$. If $V \subset A_k^N$ and $W \subset A_k^M$ are algebraic sets and if we let $\mathfrak{a}$ be the ideal in $k[X_1, \ldots, X_N, Y_1, \ldots, Y_M]$ generated by $I(V) \cup I(W) = (I(V), I(W))$, then $V(\mathfrak{a})$ is an affine algebraic set.

Proposition 12.2. $V(\mathfrak{a}) = V \times W$ (*as sets*).

Proof: We have $V = \{\xi : f_i(\xi) = 0, f_i \in I(V)\}$, $W = \{\eta : g_j(\eta) = 0, g_j \in I(W)\}$ so that clearly $V \times W \subset V(\mathfrak{a})$. On the other hand, if $(\xi, \eta) \in V(\mathfrak{a})$, then $f_i(\xi) = 0$ and $g_j(\eta) = 0$ so that $\xi \in V$ and $\eta \in W$.

Definition 12.3. $V(\mathfrak{a})$ is called the *product of V and W* and shall be denoted by $V \times W$.

Since $I(V(\mathfrak{a})) = \sqrt{\mathfrak{a}}$, the affine coordinate ring $k[V \times W]$ is given by $k[X, Y]/\sqrt{\mathfrak{a}} = k[X_1, \ldots, X_N, Y_1, \ldots, Y_M]/\sqrt{\mathfrak{a}}$. We now have, in view of the properties of tensor products (appendix A), the following:

Theorem 12.4. $k[V \times W] = k[V] \otimes_k k[W]$.

Proof 1. Let us, by abuse of notation, let $\overline{X} = (\overline{X}_1, \ldots, \overline{X}_N), \overline{Y} =$

$(\overline{Y}_1, \ldots, \overline{Y}_N)$ be the appropriate residue class (e.g. $k[V] = k[\overline{X}] = k[\overline{X}_1, \ldots, \overline{X}_N]$, $\overline{X}_i$ being the $I(V)$-residue of X_i, and $k[V \times W] = k[\overline{X}, \overline{Y}] = k[\overline{X}_1, \ldots, \overline{X}_N, \overline{Y}_1, \ldots, \overline{Y}_N]$, $\overline{X}_i$ being the $I(V \times W) = \sqrt{a}$-residue of X_i). Consider the k-homomorphism $\psi : k[V] \otimes_k k[W] \to k[V \times W]$ given by

$$\psi(\Sigma f_i(\overline{X}) \otimes g_j(\overline{Y})) = \Sigma f_i(\overline{X}) g_j(\overline{Y}) \tag{12.5}$$

Clearly ψ maps $k[V] \otimes_k k[W]$ into $k[V \times W]$. ψ is surjective since $\psi(\overline{X}_i \otimes 1) = \overline{X}_i$, $\psi(1 \otimes \overline{Y}_j) = \overline{Y}_j$ and these generate the ring $k[V \times W]$. To show that ψ is injective it will be enough to show that if $f_i(\overline{X})$, $g_j(\overline{Y})$ are independent over k, then $\psi(f_i \otimes g_j)$ are independent over k. But, if $\Sigma c_{ij} f_i(v) g_j(w) = 0$, $c_{ij} \in k$, for all v, w, then, for fixed v, $\Sigma c_{ij} f_i(v) g_j(\cdot) = 0$ on W. This implies that $\Sigma_i c_{ij} f_i(\cdot) = 0$ on V and so, $c_{ij} = 0$ for all i, j.

Proof 2. Clearly $k[V] \otimes_k k[W] = (k[X]/I(V)) \otimes_k (k[Y]/I(W))$. From appendix A, we have that $k[X] \otimes_k k[Y] = k[X, Y]$ and that $(k[X]/I(V)) \otimes_k (k[Y]/I(W)) = k[X, Y]/(I(V) \cdot k[Y] + k[X] \cdot I(W))$. Thus there is a natural k-homomorphism $\psi : k[V] \otimes_k k[W] \to k[V \times W]$ given as follows: if $f \in k[V]$ and $g \in k[W]$, then f is the restriction of a polynomial $F(X)$ to V and g is the restriction of a polynomial $G(Y)$ to W; we let $\psi(f \otimes g)$ be the restriction of $F(X)G(Y)$ to $V \times W$. Since ψ is obviously surjective, we need only show that ψ is injective. If $\sum_{i=1}^{r} f_i \otimes g_i \neq 0$ is an element of the kernel of ψ with r as small as possible ($f_i \otimes g_i \neq 0$ all i), then there is a $v \in V$ with $f_i(v) \in k$ and not all 0. It follows that $\sum_{i=1}^{r} f_i(v) g_i(\cdot) = 0$ on W and hence that (say) $g_1 = \sum_{j=2}^{r} c_j g_j$. But then $\sum_{j=2}^{r} (f_j + c_j f_1) \otimes g_j$ is an element of the kernel of ψ with fewer than r terms which is a contradiction.

Corollary 12.6. *If V and W are irreducible, then $V \times W$ is irreducible.*

Proof 1 (Appendix A): Since $k[V]$ and $k[W]$ are integral domains, so is $k[V] \otimes_k k[W]$.

Proof 2 (direct). Suppose that $V \times W = Z_1 \cup Z_2$. Then $V = V_1 \cup V_2$ where $V_i = \{v : \{v\} \times W \subset Z_i\}$ for $i = 1, 2$ (since $\{v\} \times W = (\{v\} \times W) \cap Z_1 \cup (\{v\} \times W) \cap Z_2$). But each V_i is closed. For, say, let $V_w^1 = \{v : (v, w) \in Z_1\} \simeq V \times \{w\} \cap Z_1$. Then V_w^1 is closed and $v \in V_1$

if and only if $(v, w) \in Z_1$ for all $w \in W$ i.e. if and only if $v \in \cap_{w \in W} V_w^1$. Thus, V_1, V_2 are closed and so, $V = V_1$ (say) and $V \times W = Z_1$.

Corollary 12.7. *Suppose that V, W are irreducible and that $\xi \in V$ $\eta \in W$. Then*

$$\mathcal{O}_{(\xi,\eta), V \times W} = (\mathcal{O}_{\xi,V} \otimes_k \mathcal{O}_{\eta,W})_{\mathfrak{m}_\xi \mathcal{O}_{\eta,W} + \mathfrak{m}_\eta \mathcal{O}_{\xi,V}} \qquad (12.8)$$

i.e. the local ring of $V \times W$ at (ξ, η) is the "localization" of the tensor product of the local ring of V at ξ and the local ring of W at η with respect to the maximal ideal $\mathfrak{m}_\xi \mathcal{O}_{\eta,W} + \mathfrak{m}_\eta \mathcal{O}_{\xi,V}$ (see section 9).

Proof: By definition,

$$\mathcal{O}_{(\xi,\eta), V \times W} = (k[V \times W])_{\mathfrak{m}_{(\xi,\eta)}} = (k[V] \otimes_k k[W])_{\mathfrak{m}_{(\xi,\eta)}}.$$

Since $k[V] \otimes_k k[W] \subset \mathcal{O}_{\xi,V} \otimes_k \mathcal{O}_{\eta,W} \subset \mathcal{O}_{(\xi,\eta), V \times W}$, it will be enough to show that the maximal ideal $\mathfrak{n}_{(\xi,\eta)}$ of functions in $\mathcal{O}_{\xi,V} \otimes_k \mathcal{O}_{\eta,W}$ which vanish at (ξ, η) is precisely the ideal $\mathfrak{m}_\xi \mathcal{O}_{\eta,W} + \mathfrak{m}_\eta \mathcal{O}_{\xi,V} = \mathfrak{m}$. Clearly $\mathfrak{m} \subset \mathfrak{n}_{(\xi,\eta)}$. On the other hand, if $\Sigma(f_i \otimes g_i) \in \mathcal{O}_{\xi,V} \otimes_k \mathcal{O}_{\eta,W}$ and $f_i(\xi) = a_i, g_i(\eta) = b_i$, then $\Sigma(f_i \otimes g_i) - \Sigma(a_i \otimes b_i) = \Sigma(f_i - a_i) \otimes g_i + \Sigma a_i \otimes (g_i - b_i) \in \mathfrak{m}$. So, if $\Sigma(f_i \otimes g_i)(\xi, \eta) = \Sigma f_i(\xi) \otimes g_i(\eta) = \Sigma f_i(\xi) g_i(\eta) = \Sigma a_i b_i = \Sigma a_i \otimes b_i = 0$, then $\Sigma(f_i \otimes g_i) \in \mathfrak{m}$.

Definition 12.9.   Let $\pi_V : V \times W \to V$ be given by $\pi_V(\xi, \eta) = \xi$ and $\pi_W : V \times W \to W$ be given by $\pi_W(\xi, \eta) = \eta$. Then π_V is the *projection of $V \times W$ on V* and π_W is the *projection of $V \times W$ on W*.

We observe that the projections are morphisms since, for example, π_V is the restriction to V of the mapping $\pi_1 : \mathbb{A}_k^N \times \mathbb{A}_k^M \to \mathbb{A}_k^M$ given by

$$\pi_1(x_1, \ldots, x_N, y_1, \ldots, y_M) = (x_1, \ldots, x_N) \qquad (12.10)$$

which is clearly a polynomial map. We note that the comorphism π_1^* is the natural injection of $k[x_1, \ldots, x_N] \to k[x_1, \ldots, x_N, y_1, \ldots, y_M]$, and that π_V^* is the natural injection of $k[V] \to k[V] \otimes_k k[W]$. Clearly π_V is a surjective morphism.

The product is "categorical" in that given an algebraic set X and morphisms $\phi : X \to V$, $\psi : X \to W$, there is a unique morphism $\phi \times \psi : X \to V \times W$ such that $\phi = \pi_V \circ (\phi \times \psi)$ and $\psi = \pi_W \circ (\phi \times \psi)$.

Note that, necessarily, $(\phi \times \psi)(x) = (\phi(x), \psi(x))$. Put another way, the diagram

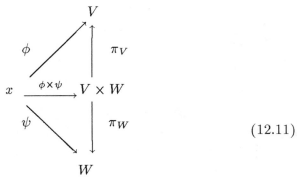

$$(12.11)$$

is a commutative diagram of morphisms. It follows that a base for the topology in $V \times W$ is given by the principal affine open sets $(V \times W)_f$ where $f(x,y) = \Sigma f_i(x)g_i(y)$ with f_i in $k[V]$ and g_i in $k[W]$. We use this observation to prove the following proposition which will be useful in section 17.

Proposition 12.12.   *The projection π_W is an open mapping.*

Proof: Let U be an open subset of $V \times W$. Then $U = \cup(V \times W)_{f_j}$ with $f_j \in k[V \times W]$. Since $\pi_V(U) = \cup \pi_V[(V \times W)_{f_j}]$, we may suppose that $U = (V \times W)_f = \{(v,w) : f(v,w) \neq 0\}$ is a principal affine open subset of $V \times W$. Let $F \in k[X,Y]$ such that F restricted to $V \times W$ is f and let (ξ_0, η_0) be an element of U. It will be enough to show that there is a U_0 open in V with $\xi_0 \in U_0$ and $U_0 \subset \pi_V(U)$. Let $G(X) = F(X, \eta_0)$ and let g be the restriction of G to V. Since $G(\xi_0) = f(\xi_0, \eta_0) \neq 0$, $G(X) \not\equiv 0$ and $g(\xi_0) \neq 0$. Thus, $U_0 = V_g$ is open in V and $\xi_0 \in U_0$. If $\xi \in U_0$, then $g(\xi) = F(\xi, \eta_0) \neq 0$ so that $(\xi, \eta_0) \in (V \times W)_f = U$ and $\xi \in \pi_V(U)$.

Proposition 12.13.   *Let $V \subset \mathbb{A}_k^N$, $W \subset \mathbb{A}_k^M$ be affine varieties and let $\phi : V \to W$, $\psi : V \to W$ be morphisms. Then the set $\{\xi : \xi \in V, \phi(\xi) = \psi(\xi)\}$ is closed.*

Proof: Let ϕ be the restriction of a polynomial map F_ϕ and let ψ be the restriction of a polynomial map F_ψ. Then $F_\phi : \mathbb{A}_k^N \to \mathbb{A}_k^M$, $F_\psi :$

$\mathbb{A}_k^N \to \mathbb{A}_k^M$ and $\{\xi : \xi \in V, \phi(\xi) = \psi(\xi)\} = V \cap \{\xi : (F_\phi - F_\psi)(\xi) = 0\}$ is closed.

Corollary 12.14. *Let* $\Delta : \mathbb{A}_k^N \to \mathbb{A}_k^N \times \mathbb{A}_k^N$ *be given by* $\Delta(\xi) = (\xi, \xi)$. *If* V *is an algebraic set, then* $\Delta(V)$ *is closed.*

Proof: Simply note that $\Delta(V) = (V \times V) \cap \{(\xi, \eta) : X_i - X_{N+i} = 0, i = 1, \ldots, N\}$.

The map Δ is called the *diagonal morphism* and $\Delta(V)$ is the diagonal of V.

Corollary 12.15. *Let* $\phi : V \to W$ *be a morphism and let* $gr(\phi) = \{(\xi, \eta) \in V \times W : \eta = \phi(\xi)\}$. *Then* $gr(\phi)$ *is closed and* $gr(\phi)$ *is isomorphic to* V.

Proof: Consider the morphism $\psi : V \times W \to W \times W$ given by $\psi(\xi, \eta) = (\phi(\xi), \eta)$. Then $gr(\phi) = \psi^{-1}(\Delta(W))$ is closed. Let $\tilde{\phi} : V \to gr(\phi)$ be the morphism given by $\tilde{\phi}(\xi) = (\xi, \phi(\xi))$. Then $\tilde{\phi}$ is bijective. If $\tilde{\pi}_V : gr(\phi) \to V$ is the restriction of the projection on V to $gr(\phi)$, then $\tilde{\pi}_V$ is a bijective morphism and $(\tilde{\pi}_V \circ \tilde{\phi})(\xi) = \xi$, $(\tilde{\phi} \circ \tilde{\pi}_V)(\xi, \phi(\xi)) = (\xi, \phi(\xi))$ so that $\tilde{\pi}_V = \tilde{\phi}^{-1}$.

Corollary 12.16. *Let* $\phi : V \to W$, $\psi : V \to W$ *be morphisms. If* $\phi = \psi$ *on a dense subset of* V, *then* $\phi = \psi$ *on* V.

Corollary 12.17. *Let* $\phi : V \to W$ *be a morphism. Then there is an injective morphism* i *and a surjective morphism* π *such that* $\phi = \pi \circ i$.

Proof: Let $\pi = \pi_W : V \times W \to W$ and let $i : V \to V \times W$ be given by $i(\xi) = (\xi, \phi(\xi))$ (so that $i(V) = gr(\phi)$). Clearly $\pi \circ i = \phi$.

We have defined, in a natural way, the product $V \times W$ of any two affine algebraic sets and noted that the coordinate ring $k[V \times W]$ of the product is simply the tensor product $k[V] \otimes_k k[W]$ of the coordinate rings of the factors. We now have:

Proposition 12.18. *If* V *and* W *are irreducible, and* $f \in k[V]$, *then* $V_f \times W$ *is a quasi-affine variety with* $k[V_f \times W] \cong k[V]_f \otimes_k k[W]$.

Proof: Clearly, $V_f \times W = (V \times W)_f$ (viewing $k[X_1, \ldots, X_N]$ as a subring of $k[X_1, \ldots, X_N, Y_1, \ldots, Y_M]$) and $k[V_f \times W] = k[V \times W]_f$ (cf. section 9). Thus, it remains to show that $k[V]_f \otimes_k k[W]$ is k-isomorphic with $k[V \times W]_f$. Let $\psi : k[V_f] \otimes k[W] \to k[V \times W]_f$ be given by

$$\psi(f_1/f \otimes g) = f_1 g/f \tag{12.19}$$

where $f_1 \in k[V]$ and $g \in k[W]$. Then ψ is a k-homomorphism (by properties of the tensor product). If h/f^n is an element of $k[V \times W]_f$ with $h = \Sigma f_i g_i$, $f_i \in k[V]$, $g_i \in k[W]$, then $\psi(\Sigma f_i/f^n \otimes g_i) = h/f^n$ so that ψ is surjective. Since $k[V \times W]$ is an integral domain (corollary 12.6), $\psi(f_1/f \otimes g) = 0$ implies either $f_1 = 0$ or $g = 0$ so that $(f_1/f \otimes g) = 0$ and ψ is injective.

Example 12.20.   Let $n = 2$ and let $G = GL(2, k) = (M(2, 2; k))_{\det g} = (\mathbb{A}_k^4)_{\det g}$. Consider the product $G \times \mathbb{A}_k^8$ and note that $k[G \times \mathbb{A}_k^8] = k[\mathbb{A}_k^4 \times \mathbb{A}_k^8]_{\det g}$. If we let $G_{11}, G_{12}, G_{21}, G_{22}$ be the coordinate functions in $\mathbb{A}_k^4$ and $X_{11}, X_{12}, X_{21}, X_{22}, Y_1, Y_2, Z_1, Z_2$ be the coordinate functions in $\mathbb{A}_k^8$, then the map $\psi_G : G \times \mathbb{A}_k^8 \to \mathbb{A}_k^8$ is given by $\psi_G(g, x) = (gAg^{-1}, gb, cg^{-1})$ (cf. (11.12)) or in coordinates by

$$\frac{\begin{bmatrix} G_{11} & G_{12} \\ G_{21} & G_{22} \end{bmatrix} \begin{bmatrix} X_{11} & X_{12} \\ X_{21} & X_{22} \end{bmatrix} \begin{bmatrix} G_{22} & -G_{12} \\ -G_{21} & G_{11} \end{bmatrix}}{(G_{11}G_{22} - G_{12}G_{21})}$$

$$\begin{bmatrix} G_{11}Y_1 + G_{12}Y_2 \\ G_{21}Y_1 + G_{22}Y_2 \end{bmatrix}$$

$$[Z_1 G_{22} - Z_2 G_{21} \quad - Z_1 G_{12} + Z_2 G_{11}]/(G_{11}G_{22} - G_{12}G_{21})$$

so that ψ_G is a morphism. We observe that the comorphism $\psi_G^* : k[X, Y, Z] \to k[G, X, Y, Z]_{\det G}$ is given by $\psi_G^*(p(X, Y, Z)) = p(GX \operatorname{adj} G/\det G, GY, Z \operatorname{adj} G/\det G)$ where p is an element of $k[X, Y, Z]$. Clearly ψ_G is surjective and ψ_G^* is injective. As we have shown many times, ψ_G is not injective. Of course, all these observations hold for arbitrary n.

We shall have a number of uses for the notion of a "product" in the sequel.

13. Group Actions, Equivalence and Invariants

We now turn our attention to examining in a general way such group actions as (11.25) and the corresponding equivalence relations. The main problem consists of two parts: first, to *find* the equivalence classes under the action of a group; and second, to *parameterize* these classes via an algebraic object. (See e.g. [D-2], [F-2], [M-3], [M-4]).

Definition 13.1.   Let X be a set and let Γ be a group. Then Γ *acts on* X if there is a map $\alpha : \Gamma \times X \to X$ such that

(i) $$\alpha(\gamma_1 \gamma_2, x) = \alpha(\gamma_1, \alpha(\gamma_2, x))$$
(ii) $$\alpha(\epsilon, x) = x$$

where ϵ is the identity in Γ. The map α, which we frequently suppress, is called an *action of Γ on X*..

If $GL(X)$ is the set of endomorphisms of X^* and Γ acts on X, then the map $\mu_\gamma : X \to X$ given by $\mu_\gamma(x) = \gamma \cdot x$ is in $GL(X)$ and $\mu_\epsilon = I$, $\mu_{\gamma_1 \gamma_2} = \mu_{\gamma_1} \mu_{\gamma_2}$ so that the map $\gamma \to \mu_\gamma$ is a homomorphism of Γ into $GL(X)$.

Definition 13.2.   An element x in X is *invariant* under the action of Γ if $\gamma \cdot x = x$ for all γ in Γ and a subset $Y \subset X$ is *invariant* if $\gamma \cdot Y \subset Y$ for all γ in Γ. If $Y \subset X$, then the *stabilizer of Y*, $S(Y)$, is the set $\{\gamma \in \Gamma : \gamma \cdot Y = Y\}$.

We note that $S(Y)$ is a subgroup of Γ since $\mu_\gamma \in GL(X)$.
Let X, Y be sets and let $\mathcal{M}(X, Y) = \{f : f$ maps X into $Y\}$.

Proposition 13.3.   *If Γ acts on X and Y, then*

$$(\gamma \cdot f)(x) = \gamma f(\gamma^{-1} x) \tag{13.4}$$

defines an action of Γ on $\mathcal{M}(X, Y)$.

---

* An endomorphism is a bijective map which preserves algebraic structure.

Proof: Simply note that $(\epsilon \cdot f)(x) = \epsilon f(\epsilon x) = f(x)$ and that $[(\gamma_1 \gamma_2) \cdot f](x) = (\gamma_1 \gamma_2) f((\gamma_1 \gamma_2)^{-1} x) = \gamma_1 [\gamma_2 f(\gamma_2^{-1} \gamma_1^{-1} x)] = \gamma_1 [(\gamma_2 \cdot f)(\gamma_1^{-1} x)] = \gamma_1 \cdot [(\gamma_2 \cdot f)](x)$.

We note that $f \in \mathcal{M}(X, Y)$ is invariant under the action (13.4) if and only if $f(\gamma x) = \gamma[f(x)]$ for all γ in Γ.

Definition 13.5.   Let k be a field and let $k^* = k - \{0\}$ be the multiplicative group of units in k. A homomorphism $\chi : \Gamma \to k^*$ is called a (k)-*character* of Γ. If $f \in \mathcal{M}(X, k)$, then f is a *relative invariant of weight* χ if $f(\gamma x) = \chi(\gamma) f(x)$ for all γ and f is an *absolute invariant* if $\chi(\gamma) \equiv 1$.

If χ is a character, then Γ acts on k via $\gamma \cdot a = \chi(\gamma) a$ for $a \in k$ since $(\gamma_1 \gamma_2) a = \chi(\gamma_1 \gamma_2) a = \chi(\gamma_1) \chi(\gamma_2) a = \gamma_1 \cdot [\gamma_2 \cdot a]$. Thus f is a relative invariant of weight χ if and only if f is an invariant under this "χ-action".

Example 13.6.   Let $Q(u, v) = au^2 + 2buv + cv^2$ and let $\Gamma = GL(2, k)$. For $\gamma \in \Gamma$, set

$$\begin{bmatrix} u' \\ v' \end{bmatrix} = \gamma \begin{bmatrix} u \\ v \end{bmatrix}$$

Then $Q(u', v') = Q(\gamma_1) u^2 + 2\gamma_1' \begin{bmatrix} a & b \\ b & c \end{bmatrix} \gamma_2 uv + Q(\gamma_2) v^2 = a'u^2 + 2b'uv + c'v^2$ where the γ_i are the columns of γ. Then $b'^2 - a'c' = (\det \gamma)^2 (b^2 - ac)$ and so the discriminant $b^2 - ac$ is a relative invariant of weight $(\det \gamma)^2$. If $\Gamma = \mathcal{O}(2, k) = \{\gamma \in GL(2, k) : \gamma'\gamma = I\}$ is the orthogonal group, then, since $a' = a\gamma_{11}^2 + 2b\gamma_{11}\gamma_{21} + c\gamma_{21}^2$ and $c' = a\gamma_{12}^2 + 2b\gamma_{12}\gamma_{22} + c\gamma_{22}^2$, we have $a' + c' = a + c$ as $\gamma_{11}^2 + \gamma_{12}^2 = 1$, $\gamma_{11}\gamma_{21} + \gamma_{12}\gamma_{22} = 0$, $\gamma_{21}^2 + \gamma_{22}^2 = 1$. In other words, the "trace" of Q is an absolute invariant under $\mathcal{O}(2, k)$. Now, let $P(u, v) = Au^2 + 2Buv + Cv^2$ be a second quadratic form. Then we could seek *simultaneous invariants* of the pair (P, Q). Such an invariant would also be an invariant of the "pencil" $Q + \lambda P$. For example, if $\gamma \in GL(2, k)$, then

$$\det \begin{bmatrix} b' + \lambda B' & a' + \lambda A' \\ c' + \lambda C' & b' + \lambda B' \end{bmatrix} = (\det \gamma)^2 \det \begin{bmatrix} b + \lambda B & a + \lambda A \\ c + \lambda C & b + \lambda B \end{bmatrix}$$

using the (relative) invariance of the discriminant of $Q + \lambda P$. Equating powers of λ, we find that $a'C' + c'A' - 2b'B' = (\det \gamma)^2 (aC + cA - 2bB)$ is a (relative) simultaneous invariant.

Definition 13.7.   If Γ acts on X and $x \in X$, then the set $O_\Gamma(x) = \{y : \text{there is a } \gamma \in \Gamma \text{ with } \gamma \cdot x = y\}$ is the called the *orbit of* x *under* Γ (or *modulo* Γ).

Proposition 13.8.   *The orbits $O_\Gamma(x)$ are the equivalence classes with respect to the relation: $xE_\Gamma y$ if there is a $\gamma \in \Gamma$ with $\gamma \cdot x = y$.*

Proof: Simply note that: (1) $xE_\Gamma x$ as $\epsilon \cdot x = x$; (2) if $xE_\Gamma y$, then $\gamma \cdot x = y$ implies $\gamma^{-1} \cdot y = x$ so that $yE_\Gamma x$; and (3) if $xE_\Gamma y$ and $yE_\Gamma z$, then $\gamma_1 \cdot x = y$, $\gamma_2 \cdot y = z$ imply $(\gamma_2 \gamma_1) \cdot x = z$ so that $xE_\Gamma z$.

Lemma   13.9. *If Γ acts on X and Y and if $f \in \mathcal{M}(X, Y)$ is Γ-invariant (proposition 13.3), then*

$$f(O_\Gamma(x)) = O_\Gamma(f(x)) \qquad (13.10)$$

for all x.

Proof: If $y = \gamma \cdot x \in O_\Gamma(x)$, then $f(y) = f(\gamma \cdot x) = \gamma[f(x)]$ as f is Γ-invariant so that $f(y) \in O_\Gamma(f(x))$. On the other hand, if $w = \gamma[f(x)] \in O_\Gamma(f(x))$, then $w = \gamma f(x) = f(\gamma x) = f(y)$ where $y = \gamma \cdot x \in O_\Gamma(x)$.

We shall say that the action of Γ is *trivial on Y* (or Γ *acts trivially on Y*) if $\gamma \cdot y = y$ for all $\gamma \in \Gamma$ and $y \in Y$.

Corollary 13.11. *If Γ acts trivially on Y, then a Γ-invariant f is constant on orbits.*

Corollary 13.12. *If Γ acts trivially on Y and if f is constant on orbits (i.e. $f(O_\Gamma(x)) = \{f(x)\}$), then f is γ-invariant.*

The corollaries provide a basis for defining the notion of an abstract invariant.

Definition 13.13.   Let E be an equivalence relation on X (e.g. $E = E_\Gamma$ where Γ acts on X). Then $f \in \mathcal{M}(X, Y)$ (where Γ acts trivially on Y if $E = E_\Gamma$) is an *abstract invariant* if f is constant on orbits (i.e. if xEx_1, then $f(x) = f(x_1)$). An abstract invariant is *complete* if it is injective on orbits (i.e. if $f(x) = f(x_1)$, then xEx_1) and a surjective abstract invariant is called *independent*.

We observe that there is always an independent complete abstract invariant. For, let $X_E = X/E = \{O_E(x) : x \in X\}$ be the quotient space of X modulo E (if $E = E_\Gamma$, we write X_Γ for X_{E_Γ}). X_E is called the *orbit space of X modulo E*. We then have:

Proposition 13.14. *The projection $\pi_E : X \to X_E$ given by $\pi_E(x) = O_E(x)$ is an independent complete abstract invariant.*

Proof: Clearly π_E is constant on orbits and is surjective. But xEx_1 if and only if $\pi_E(x) = O_E(x) = O_E(x_1) = \pi_E(x_1)$ and so π_E is injective on orbits.

Theorem 13.15. (Universal Property).   *Let $f \in \mathcal{M}(X,Y)$ be any abstract invariant. Then there is a unique map $f' : X/E \to Y$ such that $f = f' \circ \pi_E$ i.e. the diagram*

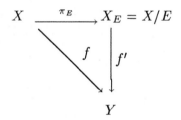

is commutative.

Proof: Simply set $f'(O_E(x)) = f(x)$ and note that f' is well-defined since f is an invariant.

Now let us return to the system theory context. Recall that the characteristic function realization map $\mathfrak{R}_\chi : A_k^{n^2+2n} \to A_k^{2n}$ given by

$$\mathfrak{R}_\chi(A, b, c) = (cb, \dots, cA^{n-1}b, -\chi_1(A), \dots, -\chi_n(A)) \qquad (13.16)$$

is a surjective morphism and that $\mathfrak{R}_\chi(S_{1,1}^n) = (A_k^{2n})_\lambda = \text{Char}(n, k)$ is isomorphic to $\text{Hank}(n, k)$ or $\text{Rat}(n, k)$. In view of proposition 11.8, the map $\mathfrak{R}_\chi$ is an invariant for the action of $G = GL(n,k)$ on $A_k^{n^2+2n}$ (given by (11.11)) and the trivial action of G on A_k^{2n}. Since $\mathfrak{R}_\chi$ is surjective, it is an independent invariant. Moreover, we have:

Proposition 13.17. *On $S_{1,1}^n$, $\mathfrak{R}_\chi$ is an independent complete abstract invariant for equivalence modulo $GL(n,k)$.*

Proof: Completeness is an immediate consequence of the State Space Isomorphism Theorem 11.19.

We observe that Proposition 13.17 remains true if we replace $\mathfrak{R}_\chi$ by either $\mathfrak{R}_h$ or $\mathfrak{R}_f$.

Let $\Sigma_{1,1}^n = S_{1,1}^n/G$ and let $\pi_G : S_{1,1}^n \to \Sigma_{1,1}^n$ be the projection given by $\pi_G(x) = O_G(x)$. Then, by the Universal Property (Theorem 13.15), there is a unique map $\varphi : \Sigma_{1,1}^n \to \mathrm{Char}(n,k)$ such that the following diagram

$$S_{1,1}^n \xrightarrow{\ \pi_G\ } \Sigma_{1,1}^n$$

with $\mathfrak{R}_\chi$ and φ

$$\mathrm{Char}(n,k)$$

is commutative. Moreover, since $\mathfrak{R}_\chi$ and π_G are both independent and complete, the mapping φ is bijective. Now $S_{1,1}^n$ and $\mathrm{Char}(n,k)$ are quasi-affine varieties and the mapping $\mathfrak{R}_\chi$ is a morphism. Since φ identifies $\Sigma_{1,1}^n$ and $\mathrm{Char}(n,k)$, we can see that, at least, the possibility of giving the orbit space an algebraic structure (e.g. that of a quasi-affine variety) exists and, hopefully, in that case the mapping φ will be an isomorphism. Of course, similar remarks apply to $\mathfrak{R}_f$ and $\mathrm{Rat}(n,k)$ and to $\mathfrak{R}_h$ and $\mathrm{Hank}(n,k)$. In order to fully develop these ideas, we turn our attention to the study of "quotients".

14. The Geometric Quotient Theorem: Introduction

Now let us try to analyze the situation from a general standpoint. We have a variety V (quasi-affine), an equivalence relation E on V which may be generated by the action of a group Γ on V, and a variety W (quasi-affine) with a map $\psi : V \to W$ which is a morphism. In the system theory case, $V = S^n_{1,1}$, $E = E_G$ where $G = GL(n,k)$, $W = \text{Char}(n,k)$ with G acting trivially on W, and $\psi = \mathfrak{R}_\chi$.

Definition 14.1. A regular function $f \in k[V]$ is an *E-invariant* (*on V*) if $v_1 E v_2$ implies $f(v_1) = f(v_2)$. The set $k[V]^E = \{f : f$ is an E-invariant$\}$ is called the *ring of E-invariants* (*on V*).

It is clear that $k[V]^E$ is a ring. In the case $E = E_\Gamma$, we write $k[V]^\Gamma$ in place of $k[V]^{E_\Gamma}$ and speak of the *ring of Γ-invariants* (or if Γ is understood from the context, simply the *ring of invariants*). We note then that $f \in k[S^n_{1,1}]^G$ if and only if

$$f^g(x) = f(g \cdot x) = f(x) \tag{14.2}$$

for all g in $G = GL(n,k)$ and $x = (A,b,c)$ in $S^n_{1,1}$. Since $k[V]^E \subset k[V]$, the ring of invariants has no nilpotent elements; however, it is not necessarily finitely generated ([N-1], [M-3], [M-4]) and so need not be an affine k-algebra. If $k[V]^E$ is finitely generated, then $k[V]^E = k[W]$ for some affine algebraic set W (proposition 7.7). However, W is not necessarily in bijective correspondence with $V_E = V/E$ ([M-4]). If so, then we can give an algebraic set structure to V_E. Thus, one critical issue is whether or not $k[V]^E$ is finitely generated. Even if this is so, we must still seek conditions which insure that $k[V]^{E} = "k[V_E]$.

Definition 14.3. An ideal $\mathfrak{a}$ in $k[V]$ is an *E-invariant ideal* if $v \in V(\mathfrak{a})$ implies that $O_E(v) \subset V(\mathfrak{a})$. We set $R = k[V]$ and $R^E =$

$k[V]^E$ so that R^E is a subring of R. If $\mathfrak{b}$ is an ideal in R^E, then $\mathfrak{a} = R\mathfrak{b}$ is an E-invariant ideal in R for: if $v \in V(\mathfrak{a})$ and $v_1 E v$, then $f \in \mathfrak{a}$ means that $f = \Sigma r_i f_i$ with $r_i \in R$, $f_i \in \mathfrak{b} \subset R^E$ so that $f(v_1) = \Sigma r_i(v_1) f_i(v_1) = \Sigma r_i(v_1) f_i(v) = 0$ (i.e. $v_1 \in V(\mathfrak{a})$). In particular, if $f \in R^E$, then Rf is an invariant ideal. If, on the other hand, $\mathfrak{a}$ is any ideal in R, then $\mathfrak{a} \cap R^E$ is an ideal in R^E. We call $\mathfrak{a} \cap R^E = \mathfrak{a}^c$ the *contraction* of $\mathfrak{a}$ to R^E and $\mathfrak{b}^e = R\mathfrak{b}$ the *extension* of $\mathfrak{b}$ to R.

In view of these observations, we have the following definitions:

Definition 14.4.   A subset Y of V is *E-invariant* if $v \in Y$ implies that $o_E(v) \subset Y$.

Definition 14.5.   The equivalence E is *proper* if (i) $k[V]^E$ is finitely generated and (ii) if $\mathfrak{a}_1, \ldots, \mathfrak{a}_r$ (arbitrary r) are invariant ideals, then $(\Sigma \mathfrak{a}_i) \cap k[V]^E = \Sigma(\mathfrak{a}_i \cap k[V]^E).$*

We note first of all that if $\mathfrak{a}$ is an invariant ideal, then $V(\mathfrak{a})$ is a closed invariant subset of V. Conversely, if Y is a closed invariant subset of V, then $I(Y)$ is an invariant ideal. If E is proper, then $k[V]^E$ is an affine k-algebra and so $k[V]^E = k[W]$ for some algebraic set W. Since $k[V]^E \subset k[V]$, there is a natural injection $\psi^* : k[V]^E \to k[V]$ which defines a natural map $\psi : V \to W$ given by

$$\psi(\alpha) = \alpha \circ \psi^*, \quad \alpha \in \operatorname{Hom}_k(k[V], k) \tag{14.6}$$

or, equivalently, in terms of maximal ideals, by

$$\psi(\mathfrak{m}) = \mathfrak{m} \cap k[W] = \mathfrak{m} \cap k[V]^E, \quad \mathfrak{m} \in \operatorname{Spm}(k[V]) \tag{14.7}$$

(See Theorem 7.5 and Corollary 7.6). It follows that ψ is constant on orbits for: if vEv_1 and $f \in k[V]^E$, then $f(v) = f(v_1)$ so that $\mathfrak{m}_v \cap k[V]^E = \mathfrak{m}_{v_1} \cap k[V]^E$. Condition (ii) of definition 14.5 can then be interpreted geometrically as follows: if V_i are closed invariant subsets of V, then

---

* This is equivalent to the seemingly more general condition: if $\{\mathfrak{a}_i\}$ is a family of invariant ideals, then $(\Sigma \mathfrak{a}_i) \cap k[V]^E = \Sigma(\mathfrak{a}_i \cap k[V]^E)$.

$$\overline{\psi(\cap V_i)} = \cap \overline{\psi(V_i)} \tag{14.8}$$

where $^{\overline{}}$ indicates closure in W. We then have:

Lemma 14.9. *If E is proper, then the natural morphism ψ maps closed invariant sets into closed sets and consequently,*

$$\psi(\cap V_i) = \cap \psi(V_i) \tag{14.10}$$

where V_i are closed invariant subsets of V.

Proof: Let V_1 be a closed invariant subset of V, let w be an element of W, and let $V_1' = \psi^{-1}(w)$. Note that V_1' is a closed invariant set (since ψ is constant on orbits and is a morphism). Then $\overline{\psi(V_1 \cap V_1')} = \overline{\psi(V_1)} \cap \{w\}$. Therefore, if w is an element of $\overline{\psi(V_1)}$, $V_1 \cap V_1'$ must be non-empty. In other words, $w \in \psi(V_1)$ and so $\psi(V_1)$ is closed.

Corollary 14.11. *If E is proper, then the natural morphism ψ is surjective. In other words, ψ is an independent invariant.*

Proof: Since ψ^* is injective, $\overline{\psi(V)} = W$ by proposition 7.18. But V is a closed invariant subset of V and so, $\psi(V)$ is closed. Thus, $\psi(V) = W$ and ψ is surjective.

Lemma 14.12. *If E is proper and $\mathfrak{a}_1$, $\mathfrak{a}_2$ are E-invariant ideals, then $V(\mathfrak{a}_1) \cap V(\mathfrak{a}_2) = \phi$ if and only if there is an f in $k[V]^E$ such that $f = 0$ on $V(\mathfrak{a}_1)$ and $f = 1$ on $V(\mathfrak{a}_2)$.*

Proof: Clearly if there is such an f, then $V(\mathfrak{a}_1) \cap V(\mathfrak{a}_2) = \phi$. If $V(\mathfrak{a}_1) \cap V(\mathfrak{a}_2) = \phi$, then, by the Nullstellensatz, $\mathfrak{a}_1 + \mathfrak{a}_2 = k[V]$. Since E is proper and the $\mathfrak{a}_i$ are invariant, $(\mathfrak{a}_1 + \mathfrak{a}_2) \cap k[V]^E = (\mathfrak{a}_1 \cap k[V]^E) + (\mathfrak{a}_2 \cap k[V]^E) = k[V]^E$. But then $1 = f_1 + f_2$ with $f_i \in \mathfrak{a}_i \cap k[V]^E$ so that $f_i(V(\mathfrak{a}_i)) = 0$ and (say) $f_2 = 1$ on $V(\mathfrak{a}_2)$.

Corollary 14.13. *If E is proper and the orbits under E are closed, then vEv_1 if and only if*

$$\psi(\mathfrak{m}_v) = \mathfrak{m}_v \cap k[V]^E = \mathfrak{m}_{v_1} \cap k[V]^E = \psi(\mathfrak{m}_{v_1}) \tag{14.14}$$

(where $m_\xi = \{f : f(\xi) = 0, f \in k[V]\}$). *In other words, ψ is a complete invariant.*

Proof: If vEv_1, then $\psi(m_v) = \psi(m_{v_1})$ since ψ is an invariant. On the other hand, since orbits are closed, $o_E(v)$ and $o_E(v_1)$ are closed invariant sets and $I(o_E(v))$ and $I(o_E(v_1))$ are invariant ideals. If $o_E(v) \cap o_E(v_1) = \phi$, then there is an f in $k[V]^E$ with $f(v) = 0$ and $f(v_1) = 1$ i.e. $f \in \psi(m_v)$ but $f \notin \psi(m_{v_1})$. It follows that $\psi(m_v) = \psi(m_{v_1})$ implies that $o_E(v) = o_E(v_1)$.

Combining corollaries 14.11 and 14.13, we have:

Corollary 14.15. *If E is proper and the orbits under E are closed, then ψ is a complete independent abstract invariant and there is a natural bijection θ_E which makes the diagram*

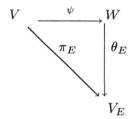

commutative.

Corollary 14.16. *If E is proper and the orbits under E are closed, then for each invariant open set $U \subset V$, there is an open set $U_0 \subset W$ such that $\psi^{-1}(U_0) = U$.*

Proof: If U is an invariant open set, then $V - U$ is an invariant closed set and $U_o = W - \psi(V - U)$ is an open set in W (by lemma 14.9) with $\psi^{-1}(U_o) \subset U$. If v_o is an element of U, then $o_E(v_o) \cap (V - U) = \phi$ (as U is invariant). Since $o_E(v_o)$ is a closed invariant set, it follows from lemma 14.9 that $\psi(o_E(v_o)) \cap \psi(V - U) = \phi$ so that $o_E(v_o) \subset \psi^{-1}(U_o)$. Thus, $\psi^{-1}(U_o) = U$.

The full import of these ideas will only become clear as we progress through the sequel.

Example 14.17.   Let $GL(2, \mathbb{C})$ act on $M(2, 2; \mathbb{C})$ by $g \cdot A = gAg^{-1}$.
If

$$A_\alpha = \begin{bmatrix} \lambda & \alpha \\ 0 & \lambda \end{bmatrix} \quad, \quad \alpha \in \mathbb{C}$$

and

$$g_\alpha = \begin{bmatrix} 1 & 0 \\ 0 & \alpha^{-1} \end{bmatrix} \quad, \quad \alpha \in \mathbb{C}^* \qquad \text{(i.e. } \alpha \neq 0)$$

then $g_\alpha \cdot A_1 = g_\alpha A_1 g_\alpha^{-1} = A_\alpha$ so that $A_\alpha \in o_G(A_1)$. But $\lim_{\alpha \to 0} A_\alpha = A_0 = \lambda I$ and $A_0 \notin o_G(A_1)$. In other words, the orbits are not closed.

Let us now define the notion of "quotient".

Definition 14.18.   ([F-2], [M-3]) Let E be an equivalence on V. A *quotient of V under E (or modulo E)* is a pair (W, ψ) consisting of an algebraic set W and a morphism $\psi : V \to W$ such that (i) ψ is an abstract E-invariant (i.e. ψ is constant on orbits), and (ii) if X is an algebraic set and $\eta : V \to X$ is an E-invariant morphism, then there is a unique morphism $\eta' : W \to X$ with $\eta = \eta' \circ \psi$. If $E = E_\Gamma$ where Γ acts on V, then Γ is assumed to act trivially on both W and X and (W, ψ) is called a *quotient of V modulo* Γ.

Clearly, the quotient is unique to within isomorphism. Moreover, we have shown that, if E is proper and orbits are closed, then (V_E, π_E) may be viewed as a quotient of V modulo E. In fact, we have actually shown even more; namely, that if E is proper and orbits are closed, then (V_E, π_E) may be viewed as a geometric quotient of V modulo E in the sense of the following:

Definition 14.19.   ([M-3]) Let E be an equivalence on V. A *geometric quotient of V modulo E* is a pair (W, ψ) consisting of an algebraic set W and a morphism ψ such that

(i) for each $w \in W$, $\psi^{-1}(w)$ is a closed orbit;

(ii) for each invariant open set $U \subset V$, there is an open set $U_o \subset W$ such that $\psi^{-1}(U_o) = U$;

(iii) the comorphism $\psi^* : k[W] \to k[V]$ is a surjective k-isomorphism between $k[W]$ and the ring of invariants $k[V]^E$.

If $E = E_\Gamma$ where Γ acts on V, then we speak of a *geometric quotient of V modulo Γ*.

We also note that, in view of the results in Appendix B, if G is a reductive group which acts on V and if the orbits under G are closed, then (V_G, π_G) may be viewed as a geometric quotient of V modulo G.

Now let us return again to the system theory context. We then have the key result:

Theorem 14.20 (Geometric Quotient Theorem). *Let $G = GL(n,k)$ act on $S_{1,1}^n$ via state equivalence. Then $(\mathrm{Char}(n,k), \mathfrak{R}_\chi)$, where $\mathfrak{R}_\chi$ is the characteristic function realization map, is a geometric quotient of $S_{1,1}^n$ modulo G.*

We shall not prove this theorem immediately since the proof takes place in stages and requires the introduction of considerable additional algebra and algebraic geometry. Also, we shall give several proofs of the various stages so that final completion of the result is a fair bit away.

Before we begin, we observe that this theorem combined with proposition 10.14 and corollary 10.15 shows that the diagram

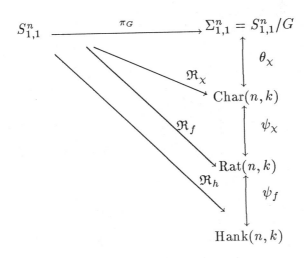

is commutative (where θ_χ is the natural bijection of theorem 13.15) and that $(\text{Rat}(n,k), \mathfrak{R}_f)$ and $(\text{Hank}(n,k), \mathfrak{R}_h)$ are also geometric quotients of $S^n_{1,1}$ modulo G. This means, of course, that the various representations of a linear system of degree n are all algebraically the same!

Now let us sketch the path that we will follow to obtain the geometric quotient theorem. We first examine the orbits giving a direct proof of closure and then giving a proof using the concept of dimension and the so-called "closed orbit lemma" which states that the orbits of minimal dimension under the action of an algebraic group are closed. We then examine the second property of a geometric quotient. The first proof we give of this property depends on the state space isomorphism theorem, the existence of minimal realizations, and dimension theoretic properties of morphisms. The second proof simply involves the observation that E_G is proper since G is reductive (Appendix B). Finally, we shall show that $k[S_{n_{1,1}}]^G = k[\text{Char}(n,k)]$ " $= $ " $k[cb, \ldots, cA^{n-1}b, \chi_1(A), \ldots, \chi_n(A)]$. The first proof is simple and depends on, in effect, a "canonical form." The generalization of this proof to the multivariable case is quite subtle since no "global" canonical form exists. The second proof we give involves invariants under the symmetric group and a good bit of algebra (Appendix C). The third proof is a translation of the methods of classical invariant theory to the system theory context and generalizes rather smoothly to the multivariable case. A fourth proof simply involves invoking the Hilbert-Mumford theory (Appendix B). While such a collection of proofs may seem overwhelming each provides a different insight into the problem.

15. The Geometric Quotient Theorem: Closed Orbits

Consider the polynomial ring $R = k[\mathbb{A}_k^{n^2+2n}]$ which we write $k[(X_{ij}), (Y_j), (Z_i)]$, $i, j = 1, \ldots, n$ or simply, $k[X, Y, Z]$. We let ψ_i (X, Y, Z) be the polynomials in R given by

$$\begin{aligned}
\psi_i(X, Y, Z) &= ZX^{i-1}Y & i &= 1, \ldots, n \\
\psi_i(X, Y, Z) &= -\chi_{i-n}(X) & i &= n+1, \ldots, 2n
\end{aligned} \tag{15.1}$$

where the χ's are the characteristic polynomials and let $\psi : \mathbb{A}_k^{n^2+2n} \to \mathbb{A}_k^{2n}$ be the regular map $(\psi_1, \ldots, \psi_{2n})$ (In other words, $\psi = \mathfrak{R}_\chi$ to simplify notation.). If $U_1, \ldots, U_{2n}$ are the coordinate functions on $\mathbb{A}_k^{2n}$ then we have $\psi^*(U_i) = \psi_i(X, Y, Z)$ and

$$k[\mathrm{Char}(n, k)] = k[U_1, \ldots, U_{2n}]_{\lambda(U_1, \ldots, U_{2n})}$$

where $\lambda(U_1, \ldots, U_{2n}) = (\theta \circ \psi_h)(U_1, \ldots, U_{2n})$ (ψ_h is given by 10.10 and θ is the Hankel determinant). If we let $S = k[\mathrm{Char}(n, k)]$, then it follows that

$$\psi^*(S) = k[\psi_1, \ldots, \psi_{2n}]_{\boldsymbol{\theta}} \tag{15.2}$$

where $\boldsymbol{\theta}(X, Y, Z) = \omega(X, Y, Z)\gamma(X, Y, Z) = \det \mathbf{H}(X, Y, Z)$ (cf. section 11, especially 11.3). We recall that $G = GL(n, k)$ acts on R via

$$g \cdot (X, Y, Z) = (gXg^{-1}, gY, Zg^{-1}) \tag{15.3}$$

and that this action preserves degrees and is by automorphisms of R. If $x_o = (A_o, b_o, c_o)$ is an element of $\mathbb{A}_k^{n^2+2n}$, then $o_G(x_o)$ is the orbit of x_o under the action of G. Let I_{x_o} be the ideal $(\psi_1(X, Y, Z) - \psi_1(A_o, b_o, c_o)), \ldots, \psi_{2n}(X, Y, Z) - \psi_{2n}(A_o, b_o, c_o))$. Then $V(I_{x_o})$ is an affine algebraic set (i.e. is closed in $\mathbb{A}_k^{n^2+2n}$) and we have:

Proposition 15.4. $o_G(x_o) \subset V(I_{x_o})$.

Proof: If $g \cdot x_o = x_1 = (A_1, b_1, c_1)$, then $A_1 = gA_og^{-1}$, $b_1 = gb_o$, $c_1 = c_og^{-1}$ and $c_oA_o^{j-1}b_o = c_1A_1^{j-1}b_1$, $\chi_i(A_1) = \chi_i(A_o)$ so that $\psi_i(A_1, b_1, c_1) = \psi_i(A_o, b_o, c_o)$.

Corollary 15.5. *If $x_o \in S_{1,1}^n$, then $o_G(x_o) = V(I_{x_o}) \cap S_{1,1}^n$.*

Proof: If $x_1 \in o_G(x_o)$, then $x_1 \in V(I_{x_o})$ by the proposition and $x_1 \in S_{1,1}^n$ by (say) corollary 11.15. On the other hand, if $x_1 \in V(I_{x_o}) \cap S_{1,1}^n$, then $\psi(x_1) = \psi(x_o)$ and $\Re_\chi(x_1) = \Re_\chi(x_o)$. It follows from the state space isomorphism theorem that $x_1 \in o_G(x_o)$.

Of course, corollary 15.5 means that *orbits in $S_{1,1}^n$ are closed.* Off $S_{1,1}^n$, there is trouble as the following examples show.

Example 15.6.   Let $n = 2$, $A_o = I_2$, $b_o = \begin{bmatrix} 0 \\ 1 \end{bmatrix}$ and $c_o = [1 \ \ 0]$. Then $c_o A_o^j b_o = 0$ for all j so that $\psi(x_o) = (0, 0, -2, 1) = \psi(x_1)$ where $A_1 = I_2$, $b_1 = \begin{bmatrix} 0 \\ 0 \end{bmatrix}$, $c_1 = [1 \ \ 0]$. But x_1 is not in $o_G(x_o)$ as $b_o \neq 0$ implies $gb_o \neq 0$ for all g in G.

Example 15.7.   Now let $n = 3$ and

$$A_o = \begin{bmatrix} 1 & 0 & 0 \\ 0 & 1 & 0 \\ 0 & 0 & 1 \end{bmatrix} \ , \quad b_o = \begin{bmatrix} 0 \\ 0 \\ 1 \end{bmatrix} \ , \quad c_0 = [1 \ \ 0 \ \ 1]$$

Then $c_o A_o^j b_o = 1$ for all j. If $x_o = (A_o, b_o, c_o)$, then we have $\psi(x_o) = (1, 1, 1, -3, 3, -1)$, $f_{x_o}(z) = 1/(z - 1)$ and rank $H_{x_o} = 1$. If

$$A_1 = \begin{bmatrix} 1 & 1 & 0 \\ 0 & 1 & 0 \\ 0 & 0 & 1 \end{bmatrix} \ , \quad b_1 = \begin{bmatrix} 1 \\ 0 \\ 0 \end{bmatrix} \ , \quad c_1 = [1 \ \ 0 \ \ 0]$$

then $c_1 A_1^j b_1 = 1$ for all j. If $x_1 = (A_1, b_1, c_1)$, then we have $\psi(x_1) = (1, 1, 1, -3, 3, -1) = \psi(x_o)$ and $f_{x_1}(z) = 1/(z - 1) = f_{x_o}(z)$ but $x_1 \notin o_G(x_o)$ since $g A_o g^{-1} = A_o \neq A_1$ for all g in G.

Now, since the orbits $o_G(x_o)$ in $S_{1,1}^n$ are closed, we have:

Proposition 15.8.   *Let ξ be an element of* $\mathrm{Char}(n, k)$. *Then $\psi^{-1}(\xi) = o_G(x)$ for some x in $S_{1,1}^n$, and hence, requirement* (i) *for a geometric quotient is satisfied.*

Proof: By theorem 10.18 on the existence of minimal realizations, $\psi^{-1}(\xi)$ is not empty and there is an $x = (A, b, c)$ in $S_{1,1}^n$ with $\psi(x) = \xi$. By the state space isomorphism theorem 11.19, if $x_1 \in \psi^{-1}(\xi)$, then there is a unique g_{x_1} in G such that $g_{x_1} \cdot x_1 = x$ and so $x_1 \in o_G(x)$. In other words, $\psi^{-1}(\xi) \subset o_G(x)$. On the other hand, if $x_2 \in o_G(x)$, then $\psi(x_2) = \psi(x) = \xi$ by proposition 13.17 so that $o_G(x) \subset \psi^{-1}(\xi)$.

Now let x_0 be an element of $S_{1,1}^n$ and let $S(x_o) = \{g : g \cdot x_o = x_o\}$ be the *stabilizer of* x_o. Then $S(x_o) = \{I\}$ by the state space isomorphism theorem. Since the mapping $G \longrightarrow G \cdot x_o = o_G(x_o)$ is a bijective morphism between $G/S(x_o)$ and $o_G(x_o)$, a reasonable notion of dimension for affine (or quasi-affine) algebraic sets would imply that the dimension of each orbit is the same as the dimension of $G/S(x_o) = G$ (i.e. dimension is invariant under isomorphism). A second reasonable requirement for the notion of dimension is that if V_1 is strictly contained in V_2 (with V_2 irreducible), then the dimension of V_1 should be strictly less than the dimension of V_2, written $\dim V_1 < \dim V_2$. We shall, in a lengthy algebraic digression, develop such a notion of dimension and show that, in fact, $\dim o_G(x_o) = \dim GL(n, k) = n^2$ for all x_o in $S_{1,1}^n$.

Assuming the requisite properties of dimension, we have the following:

Lemma 15.9 (Closed Orbit Lemma). *Orbits of minimal dimension are closed.*

Proof: Let x_o be an element of $S_{1,1}^n$ and $V = G \cdot x_o = o_G(x_o)$. Since $o_G(x_o) = V(I_{x_o}) \cap S_{1,1}^n$ is a locally closed set, it contains a dense open subset of its closure $\overline{V}$ in $\mathbb{A}_k^{n^2+2n}$. But G acts transitively on V as $G \cdot x = o_G(x) = o_G(x_o) = V$ for all x in $o_G(x_o)$. Moreover, $G \cdot \overline{V} \subset \overline{V}$ (i.e. $\overline{V}$ is stable under the action of G). It follows that V contains a neighborhood in $\overline{V}$ of each of its points. In other words, V is open in $\overline{V}$ and so, $V = \overline{V} \cap U$ with U open in $\mathbb{A}_k^{n^2+2n}$. Therefore, $\overline{V} - V = \overline{V} - (\overline{V} \cap U) = W \cap \overline{V}$ is closed where $W = \mathbb{A}_k^{n^2+2n} - U$ is the *closed* complement of U. If $\overline{V} - V$ is not empty, then $\overline{V} - V$ is strictly contained in $\overline{V}$ and $\dim(\overline{V} - V) < \dim \overline{V}$. Since $\overline{V} - V$ is stable under the action of G (i.e. $G \cdot (\overline{V} - V) \subset \overline{V} - V$), $\overline{V} - V$ contains orbits of lower dimension. Thus, if $V = o_G(x_o)$ were an orbit of minimal dimension, then $\overline{V} - V$ would be empty and so $V = \overline{V}$ would be closed.

We note that lemma 15.9 holds in the more general context of an algebraic group acting on a variety ([H-8]).

16. Affine Algebraic Geometry: Dimension

So let us now turn our attention to developing the notion of dimension. We begin with the definition of topological dimension. Then, we introduce the transcendence degree and use it to define the dimension of a prime ideal. This provides an algebraic notion of dimension. An analysis of integral dependence follows and culminates in the "Going Down" theorem. This allows us to relate the dimension of a prime ideal to its depth and height. We can then show the equivalence of the topological and algebraic definitions of dimension. This in turn leads us to the result that $\dim_k GL(n, k) = n^2$. We conclude with some results on varieties of codimension r which will be useful in treating fibers of morphisms. Of course, the lengthy development will be quite useful throughout the sequel.*

Definition 16.1. If X is a topological space, then the (topological) *dimension of* X, $\dim X$, is the maximum integer n such that there is a chain $V_0 < V_1 \cdots < V_n$ of distinct closed irreducible subsets of X. If $V \subset \mathbb{A}_k^N$ is an affine or quasi-affine variety, then the (topological) *dimension of* V, $\dim_k V$, is its dimension as a topological space (with the Zariski topology). If $V \subset \mathbb{A}_k^N$ is an affine algebraic set, then the (topological) *dimension of* V, $\dim_k V$, is the maximum of $\dim_k V_i$ where $V = \cup V_i$ is the decomposition of V into its irreducible components (proposition 6.10).

Now let $R = k[V]$ where V is an affine variety. Then R is an integral domain since $\mathfrak{p} = I(V)$ is a prime ideal. We let $K(R)$ be the quotient field of R. The elements of $K(R)$ are of the form f/g where $f, g \in k[V]$ and $g \neq 0$ and are called *rational functions on* V. In this situation, we write $K(V)$ in place of $K(k[V]) = K(R)$ and call $K(V)$ the *function field of* V (see section 9).

We now make the following:

Definition 16.2. Let K be an extension field of k. Elements $\xi_1, \ldots, \xi_N$ of K are *algebraically independent over* k if $f \in k[x_1, \ldots,$

---

* Our treatment relies on [A-2], [D-1], [H-2], [H-8], [M-1] and [Z-3].

x_N], $f(\xi_1, \ldots, \xi_N) = 0$ implies $f \equiv 0$. A maximal set of algebraically independent elements is called a *transcendence basis of K over k*. If $K = k(z_1, \ldots, z_N)$ is finitely generated, then the maximum number (finite) of algebraically independent elements is called the *transcendence degree of K over k* and is written tr. deg K/k. If $R = k[z_1, \ldots, z_N]$ is a finitely generated integral domain with quotient field $K(R) = (k(z_1, \ldots, z_N))$, then the *transcendence degree of R over k*, tr. deg R/k, is tr. deg $K(R)/k$. If $\mathfrak{p}$ is a prime ideal in R, then the *dimension of $\mathfrak{p}$ over k*, $\dim_k \mathfrak{p}$, is the transcendence degree of $R/\mathfrak{p}$ over k i.e. $\dim_k \mathfrak{p} = $ tr. $\deg(R/\mathfrak{p})/k = $ tr. deg $K(R/\mathfrak{p})/k$. In particular, if $R = k[\mathbb{A}_k^N] = k[X_1, \ldots, X_N]$ and $\mathfrak{p} = I(V)$ is the ideal of an affine or quasi-affine variety V, then the *algebraic dimension of V over k*, written $\mathrm{Dim}_k V$, is $\dim_k \mathfrak{p} = \dim_k I(V)$.

We note that

$$\mathrm{Dim}_k V = \dim_k I(V) = \text{tr. deg } k[V]/k = \text{tr. deg } K(V)/k \qquad (16.3)$$

for an affine (or quasi-affine) variety V. In other words, the algebraic dimension of V is the transcendence degree of the function field $K(V)$ of V.

Now we have defined both a topological and an algebraic dimension for V and our goal will be to show that these are the same. Before doing so, we develop some simple results which indicate that we are on the right track.

Proposition 16.4. $\mathrm{Dim}_k \mathbb{A}_k^N = N$.

Proof: $k[\mathbb{A}_k^N] = k[X_1, \ldots, X_N]$ has transcendence degree N.

Corollary 16.5. $\mathrm{Dim}_k GL(n, k) = n^2$.

Proof: The quasi-affine variety $GL(n, k)$ is open and dense in $M(n, n; k)$ so that $\mathrm{Dim}_k GL(n, k) = \mathrm{Dim}_k M(n, n; k) = \mathrm{Dim}_k \mathbb{A}_k^{n^2} = n^2$.

Corollary 16.6. $\mathrm{Dim}_k \mathrm{Rat}(n, k) = 2n$, $\mathrm{Dim}_k \mathrm{Hank}(n, k) = 2n$, *and* $\mathrm{Dim}_k \mathrm{Char}(n, k) = 2n$.

Lemma 16.7. *Let* $R' = k[y_1, \ldots, y_m]$ *be a finitely generated integral* *domain and let R be an integral domain over k. Let $\varphi : R \to R'$ be a surjective k-homomorphism. Then* tr. deg $R'/k \leq$ tr. deg R/k.

Proof: Let $r'_1, \ldots, r'_s$ be algebraically independent elements of R' and let $r_1, \ldots, r_s$ be elements of R with $\varphi(r_i) = r'_i$. If $f \in k[X_1, \ldots, X_s]$ and $f(r_1, \ldots, r_s) = 0$, then $f(\varphi(r_1), \ldots, \varphi(r_s)) = f(r'_1, \ldots, r'_s) = 0$ implies $f \equiv 0$ so that the r_i are algebraically independent.

Corollary 16.8. *If V and W are k-isomorphic varieties then* $\mathrm{Dim}_k V = \mathrm{Dim}_k W$.

Proof: Simply note that $k[V]$ and $k[W]$ are k-isomorphic.

Corollary 16.9. *If* tr. deg $R'/k =$ tr. deg R/k, *then φ is a k-isomorphism.*

Proof: Enough to show φ is injective. Let $u \in R$ with $u \neq 0$. If $r'_1, \ldots, r'_\nu$ is a transcendence basis of R'/k and if $r_1, \ldots, r_\nu$ are elements of R with $\varphi(r_i) = r'_i$, then $r_1, \ldots, r_\nu$ is a transcendence basis of R/k (being algebraically independent by the proof of the lemma). Then u is algebraic over $k[r_1, \ldots, r_\nu]$ and there is an equation

$$a_0(r_1, \ldots, r_\nu)u^t + \cdots + a_t(r_1, \ldots, r_\nu) = 0$$

with $a_i(\cdot) \in k[x_1, \ldots, x_\nu]$, $a_0(r_1, \ldots, r_\nu) \neq 0$ and $t \geq 1$ minimal (so that $a_t(r_1, \ldots, r_\nu) \neq 0$ and $a_t(x_1, \ldots, x_\nu) \not\equiv 0$). Applying φ, we get

$$a_0(r'_1, \ldots, r'_\nu)[\varphi(u)]^t + \cdots + a_t(r'_1, \ldots, r'_\nu) = 0$$

If $\varphi(u) = 0$, then $a_t(r'_1, \ldots, r'_\nu) = 0$ which would imply the contradiction $a_t(x_1, \ldots, x_\nu) \equiv 0$. Thus, $\varphi(u) \neq 0$.

Corollary 16.10. *If $\mathfrak{p}' < \mathfrak{p}$, then $\dim_k \mathfrak{p}' > \dim_k \mathfrak{p}$. In particular, if V and W are affine varieties and $V < W$, then $\mathrm{Dim}_k V < \mathrm{Dim}_k W$.*

Proof: The k-homomorphism $\varphi : R/\mathfrak{p}' \to R/\mathfrak{p} = (R/\mathfrak{p}')/(\mathfrak{p}/\mathfrak{p}')$ is surjective but is not an isomorphism.

Corollary 16.11. *Let R be a finitely generated integral domain over k and let $\mathfrak{p} \neq (0)$ be a prime ideal in R. Then $\dim_k(0) =$ tr. deg $R/k = \nu$ and $\dim_k \mathfrak{p} < \nu$.*

Corollary 16.12. $\dim_k \mathfrak{p} = 0$ *if and only if* $\mathfrak{p}$ *is maximal. In particular, if* V *is a variety, then* $\text{Dim}_k V = 0$ *if and only if* $V = \xi$ *is a point.*

Proof: If $\mathfrak{p} = \mathfrak{m}$ is maximal, then $R/\mathfrak{m} = k[z_1,\ldots,z_m]$ is a field and the z_i are algebraic over k (by lemma 5.13). On the other hand, suppose $\dim_k \mathfrak{p} = 0$. Then $\mathfrak{p} \subseteq \mathfrak{m}$ where $\mathfrak{m}$ is maximal. If $\mathfrak{p} < \mathfrak{m}$, then $\dim_k \mathfrak{p} > \dim_k \mathfrak{m} = 0$, a contradiction.

Corollary 16.13. *If* $\dim_k \mathfrak{p} = \nu - 1$, *then* $\mathfrak{p}$ *is a minimal prime ideal (i.e. if* $\mathfrak{p}' < \mathfrak{p}$, *then* $\mathfrak{p}' = (0)$). *In particular, if* $V \subset \mathbb{A}_k^N$ *and* $\text{Dim}_k V = \nu - 1$, *then* $V = V(f)$ *where* f *is irreducible in* $k[X_1,\ldots,X_N]$.*

Proposition 16.14.   *Let* $V \subset \mathbb{A}_k^N$ *and* $W \subset \mathbb{A}_k^M$ *be affine (or quasi-affine) varieties. Then* $\text{Dim}_k(V \times W) = \text{Dim}_k V + \text{Dim}_k W$.

Proof: Let $n = \text{Dim}_k V$ and $m = \text{Dim}_k W$. Then $k[V]$ contains a subalgebra $k[x_1,\ldots,x_n]$ isomorphic to a polynomial ring and $k[W]$ contains a subalgebra $k[y_1,\ldots,y_m]$ isomorphic to a polynomial ring. Moreover,

$$k[V] = (k[x_1,\ldots,x_n])[f_1,\ldots,f_s]$$

and

$$k[W] = (k[y_1,\ldots,y_m])[g_1,\ldots,g_t]$$

where the f_i are algebraic over $k(x_1,\ldots,x_n)$ and the g_j are algebraic over $k(y_1,\ldots,y_m)$. Now $k[V \times W] = k[V] \otimes_k k[W]$ (theorem 12.4) contains $k[x_1,\ldots,x_n] \otimes k[y_1,\ldots,y_m] = k[x_1,\ldots,x_n,y_1,\ldots,y_m]$ and so, $K(V \times W)$ contains $k(x_1,\ldots,x_n,y_1,\ldots,y_m)$ which has transcendence degree $n + m$ over k. But $K(V \times W)$ is an extension of $k(x_1,\ldots,x_n,y_1,\ldots,y_m)$ generated by $f_i \otimes 1$, $1 \otimes g_j$ which are algebraic over this subfield. It follows that tr. deg $K(V \times W)/k =$ tr. deg $k(x_1,\ldots,x_n,y_1,\ldots,y_m)/k = n + m$ and hence that

$$\text{Dim}_k V \times W = \text{Dim}_k V + \text{Dim}_k W.$$

Suppose V is a variety (i.e. is irreducible) and that the topological dimension of V, $\dim_k V$, is n. Then there is a chain $V_0 < V_1 < \cdots <$

---

* Note V is a variety here (i.e. is irreducible).

$V_n = V$ of closed irreducible subsets of V and a corresponding chain $I(V) < I(V_{n-1}) < \cdots < I(V_0) < k[x_1,\ldots,x_n]$. Since V_0 must be a point, $I(V_0)$ is a maximal ideal. It follows from corollaries 16.10 and 16.12 that $\mathrm{Dim}_k V > \mathrm{Dim}_k V_{n-1} > \cdots > \mathrm{Dim}_k V_0 = 0$. In other words, $\mathrm{Dim}_k V \geq n$. On the other hand, if there is a chain of prime ideals, $I(V) < \mathfrak{p}_{d-1} < \cdots < \mathfrak{p}_0 < k[x_1,\ldots,x_n]$, then $V(\mathfrak{p}_0) < \cdots < V(\mathfrak{p}_{d-1}) < V$ is a chain of closed irreducible subsets of V and $\dim_k V$ is equal to the length of a maximal chain of prime ideals containing $I(V)$. This leads us to make the following:

Definition 16.15. Let R be an integral domain and let $\mathfrak{p}$ be a prime ideal in R. Then $h = h(\mathfrak{p})$ is the *height of* $\mathfrak{p}$ if there is a (strict) chain $(0) = \mathfrak{p}_0 < \mathfrak{p}_1 < \cdots < \mathfrak{p}_h = \mathfrak{p}$ of prime ideals and there does not exist such a chain with more than $h+1$ elements. Also, $d = d(\mathfrak{p})$ is the *depth of* $\mathfrak{p}$ if there is a (strict) chain $\mathfrak{p} = \mathfrak{p}_d < \mathfrak{p}_{d-1} < \cdots < \mathfrak{p}_0 < R$ of prime ideals and there does not exist such a chain with more than $d + 1$ elements.

The key result which makes things work is a theorem which states that, in a finitely generated integral domain R over k, if $\dim_k \mathfrak{p} = s$, then $d(\mathfrak{p}) = s$ and $h(\mathfrak{p}) = \nu - s$ where $\nu = \mathrm{tr. \ deg} \, R/k$. This theorem rests on a lemma which is a converse of corollary 16.13. In order to prove the lemma we need the concept of integral dependence and some of its properties.

Definition 16.16. Let $R \subset S$ be rings. An element u of S is *integral over* R if u satisfies a monic equation

$$u^n + r_1 u^{n-1} + \cdots + r_n = 0 \tag{16.17}$$

with coefficients r_i in R.

Definition 16.18 (cf. A.16). Let R be a ring. An abelian group M on which R acts linearly is called an *R-module* (i.e. M is an R-"vector" space). More explicitly, there is a mapping $R \times M \to M, (r, m) \to rm$, such that $r(m_1 + m_2) = rm_1 + rm_2, (r + r_1)m = rm + r_1 m, (rr_1)m = r(r_1 m)$, and $1m = m$. M is a *finite* (or *finitely generated*) R-module if there are $m_1,\ldots,m_t$ in M such that $m \in M$ implies $m = \sum_{i=1}^{t} r_i m_i$ with $r_i \in R$. M is a *faithful* R-module if $rM = 0$ implies $r = 0$.

Proposition 16.19.   *Let $R \subset S$ be rings. Then the following are equivalent: (i) $u \in S$ is integral over R; (ii) $R[u]$ is a finite R-module; (iii) $R[u] \subset S_1$ a subring of S and S_1 is a finite R-module; and, (iv) there is a faithful $R[u]$-module M which is a finite R-module.*

Proof: Clearly (i) implies (ii) since $u^{n+r} = -(r_1 u^{n+r-1} + \cdots + r_n u^r)$ for $r \geq 0$ means $R[u]$ is generated over R by $1, u, \ldots, u^{n-1}$. As for (ii) implying (iii), simply let $S_1 = R[u]$. If (iii) holds, let $M = S_1$ and note that $p(u)M = 0$ implies $p(u) \cdot 1 = 0$ where $p(u) \in R[u]$ (i.e. S_1 is a faithful $R[u]$-module). Suppose that (iv) holds for M and let ϕ_u be the map of M into itself given by

$$\phi_u(m) = um \tag{16.20}$$

If $m_1, \ldots, m_t$ are generators of M over R, then

$$\phi_u(m_i) = um_i = \Sigma r_{ij} m_j \ , \quad r_{ij} \in R \tag{16.21}$$

so that

$$\Sigma(\delta_{ij} u - r_{ij}) m_j = 0 \tag{16.22}$$

for $i = 1, \ldots, t$. Multiplying by the adjoint of $(\delta_{ij} u - r_{ij})$, we have $\det(\delta_{ij} u - r_{ij})M = 0$. Since M is a faithful $R[u]$-module, $\det(\delta_{ij} u - r_{ij}) = 0$ which is an equation of integral dependence for u over R.

Corollary 16.23. *If $u_1, \ldots, u_n$ are integral over R, then the ring $R[u_1, \ldots, u_n]$ is a finite R-module.*

Proof: Induction on n.

Corollary 16.24. *Let $S_1 = \{u \in S : u$ is integral over $R\}$. Then S_1 is a subring of S.*

Proof: If $u_1, u_2 \in S_1$, then $R[u_1, u_2]$ is a finite R-module and so, $u_1 + u_2$, $u_1 u_2$ are integral over R.

Definition 16.25. $S_1 = \{u \in S; u$ is integral over $R\}$ is the *integral closure of R in S*. If $S_1 = R$, then R is *integrally closed in S*, and if $S_1 = S$, then S is *integral over R*. If R is an integral domain, then the *integral closure of* R is its integral closure in $K(R)$ (the quotient field of R) and R is *integrally closed* if R equals its integral closure.

Corollary 16.26 (Transitivity of Integral Dependence). *Let* $R \subset S \subset T$ *be rings with* S *integral over* R *and* T *integral over* S. *Then* T *is integral over* R.

Proof: If $u \in T$, then $u^n + s_1 u^{n-1} + \cdots + s_n = 0$ with $s_i \in S$. Then $R[s_1, \ldots, s_n]$ is a finite R-module and $R[s_1, \ldots, s_n][u]$ is a finite $R[s_1, \ldots, s_n]$-module together imply that $R[s_1, \ldots, s_n][u]$ is a finite R-module.

Corollary 16.27. *The integral closure* S_1 *of* R *in* S *is integrally closed in* S.

For some examples, we have:

Proposition 16.28. *A unique factorization domain* R *is integrally closed.*

Proof: (see proposition 5.14).

Corollary 16.29. *The polynomial ring* $k[X_1, \ldots, X_N]$ *is integrally closed.*

Corollary 16.30. *The ring of integers* $\mathbb{Z}$ *is integrally closed.*

Proposition 16.31. *If* R *is integrally closed and* M *is a multiplicatively closed subset of* R *(definition 9.2), then* R_M *is integrally closed.*

Proof: Since $K(R_M) = K(R)$, let $u = r_1/r_2$, $r_i \in R$, be integral over R_M. Then

$$\left(\frac{r_1}{r_2}\right)^n + \frac{s_1}{m}\left(\frac{r_1}{r_2}\right)^{n-1} + \cdots + \frac{s_n}{m} = 0 \qquad (16.32)$$

with $s_i \in R$ and $m \in M$. Multiplying by m^n,

$$\left(\frac{mr_1}{r_2}\right)^n + s_1\left(\frac{mr_1}{r_2}\right)^{n-1} + \cdots + s_n m^{n-1} = 0 \qquad (16.33)$$

and since all the coefficients are in R, we have $(mr_1/r_2) = r' \in R$ so that $r_1/r_2 = r'/m \in R_M$.

Corollary 16.34. *If V is an affine or quasi-affine variety and $k[V]$ is integrally closed, then $\mathcal{O}_{\xi,V}$ is integrally closed for all ξ in V. Conversely, if $\mathcal{O}_{\xi,V}$ is integrally closed for all ξ in V, then $k[V]$ is integrally closed.*

Proof: For the first part, simply note that $\mathcal{O}_{\xi,V} = k[V]_{\mathfrak{m}_\xi}$. As for the second part, it follows immediately from proposition 9.11.

Corollary 16.35. *$S^n_{1,1}$, $\mathrm{Char}(n,k)$, $\mathrm{Hank}(n,k)$, $\mathrm{Rat}(n,k)$ and $GL(n,k)$ all have coordinate rings which are integrally closed.*

Now let us continue on the road to showing that $\mathrm{Dim}_k V = \dim_k V$.

Proposition 16.36.   *If S is integral over R, $\mathfrak{b}$ is an ideal in S, and $\mathfrak{a} = \mathfrak{b} \cap R$, then $S/\mathfrak{b}$ is integral over $R/\mathfrak{a}$ (where $R/\mathfrak{a} = R/R \cap \mathfrak{b}$ is viewed as a subring of $S/\mathfrak{b}$).*

Proof: If $u^n + r_1 u^{n-1} + \cdots + r_n = 0$, then $\bar{u}^n + \bar{r}_1 \bar{u}^{n-1} + \cdots + \bar{r}_n = 0$.

Let $R \subset S$ and let $\mathfrak{a}$ be an ideal in R. Then $u \in S$ is *integral over* $\mathfrak{a}$ if $u^n + a_1 u^{n-1} + \cdots + a_n = 0$ with $a_i \in \mathfrak{a}$. The *integral closure of* $\mathfrak{a}$ in S is the set $\{u \in S : u$ is integral over $\mathfrak{a}\}$.

Lemma 16.37. *Let $R \subset S$, $\mathfrak{a}$ be an ideal in R, and S_1 be the integral closure of R in S. If $\mathfrak{a}_1 = S_1\mathfrak{a}$, then $\sqrt{\mathfrak{a}_1}$ is the integral closure of $\mathfrak{a}$ in S.*

Proof: If u is integral over $\mathfrak{a}$, then $u^n + a_1 u^{n-1} + \cdots + a_n = 0$ with $a_i \in \mathfrak{a}$. Thus, $u \in S_1$ and $u^n \in S_1\mathfrak{a} = \mathfrak{a}_1$ so that $u \in \sqrt{\mathfrak{a}_1}$. Conversely, if $u \in \sqrt{\mathfrak{a}_1}$, then $u^n = \Sigma s_i a_i$ with $s_i \in S_1$ and $a_i \in \mathfrak{a}$. Since the s_i are integral over R, $M = R[s_1, \ldots, s_\mu]$ is a finite R-module. But $u^n M \subset \mathfrak{a} M$ and letting m_i be the generators of M, we have $u^n m_i = \Sigma a_{ij} m_j$, $a_{ij} \in \mathfrak{a}$ so that $\Sigma(\delta_{ij} u^n - a_{ij}) m_j = 0$. It follows that $\det(\delta_{ij} u^n - a_{ij}) = 0$ (cf. proof of proposition 16.19). In other words, u^n and hence, u, is integral over $\mathfrak{a}$.

Proposition 16.38.   *Let R, S be integral domains with $R \subset S$ and R integrally closed. If u is integral over the ideal $\mathfrak{a}$ of R, then*

u is algebraic over $K(R)$ and the minimal polynomial $f(X) = X^n + a_1 X^{n-1} + \cdots + a_n$ of u has coefficients $a_1, \ldots, a_n$ in $\sqrt{\mathfrak{a}}$.

Proof: Clearly u is algebraic over $K(R)$. Let $L \supset K(R)$ be a field containing all the roots $\xi_1, \ldots, \xi_n$ of $f(X)$. Then each ξ_i is integral over $\mathfrak{a}$ (as each ξ_i satisfies the same equation as u). But the coefficients a_i are in $K(R)[\xi_1, \ldots, \xi_n]$ and so are integral over $\mathfrak{a}$. Hence, by the lemma, the a_i are in $\sqrt{\mathfrak{a}}$.

Corollary 16.39. *Let R, S be integral domains with $R \subset S$ and R integrally closed. If u is integral over R, then u is algebraic over $K = K(R)$ and the minimal polynomial $f(X) = X^n + a_1 X^{n-1} + \cdots + a_n$ of u over K has coefficients in R. If S is integral over R and $L = K(S)$, then L is an algebraic extension of K. If L/K is a finite algebraic extension and $u \in S$, then the field polynomial (cf. exercise 47 [L-1], [Z-3]) of u over K has coefficients in R so that, in particular, $u_0 = N_{L/K}(u) \in R$ where $N_{L/K}(u)$ is the norm of u. Furthermore, if $\mathfrak{p} = \sqrt{(u)}$ (an ideal in S), then $\mathfrak{p} \cap R = \sqrt{(u_0)}$. (We shall use this observation in the proof of Krull's Theorem 16.54).*

Proof: The first part is a direct consequence of the proposition. To show that L is algebraic over K, it is enough to show that $1/v$ for v in S is algebraic over K. But v is integral over R so that $v^m + a_1 v^{m-1} + \cdots + a_m = 0$ with $a_i \in R$. Hence $0 = (1/v)^m(v^m + a_1 v^{m-1} + \cdots + a_m) = 1 + a_1(1/v)^{m-1} + \cdots + a_m(1/v)^m$ and $1/v$ is algebraic over K. Since the field polynomial of u over K is a power of the minimal polynomial, ([L-1], [Z-3]) it has coefficients in R. Let $f(X) = X^n + a_1 X^{n-1} + \cdots + a_n$ be the minimal polynomial of u over K (with $a_i \in R$). If $u_0 = N_{L/K}(u)$, then $u_0 = a_n^\nu$ for some ν. Thus, if $\mathfrak{p} = \sqrt{(u)}$ and $u^n + a_1 u^{n-1} + \cdots + a_n = 0$, then $0 = (a_n^{\nu-1} u^n + \cdots + a_n^{\nu-1} a_{n-1} u) + u_0$ and $u_0 \in \mathfrak{p}$ as $u \in \mathfrak{p}$. On the other hand, if $v \in \mathfrak{p} \cap R$, then $v \in \sqrt{(u)}$ so that $v^n = us$ with $s \in S$. Taking norms, we have $v^{n[L:K]} = N_{L/K}(v^n) = N_{L/K}(us) = N_{L/K}(u) N_{L/K}(s) = u_0 N_{L/K}(s)$ (where $[L:K]$ is the degree of L/K). But $N_{L/K}(s)$ is an element of R since S is integral over R. In other words, $v \in \sqrt{(u_0)}$.

We now turn our attention to the so-called "Going-Down" theorem. Our treatment is based on [A-2].

Theorem 16.40. *Let $R \subset S$ be integral domains with R integrally closed and S integral over R. Let $\mathfrak{p}_n \subset \cdots \subset \mathfrak{p}_1$ be a chain of prime ideals in R and let $\mathfrak{q}_m \subset \cdots \subset \mathfrak{q}_1$, $m < n$, be a chain of prime ideals in S with $\mathfrak{q}_i \cap R = \mathfrak{p}_i$, $i = 1, \ldots, m$. Then the chain of $\mathfrak{q}_i$ can be extended to a chain $\mathfrak{q}_n \subset \cdots \subset \mathfrak{q}_m \subset \cdots \subset \mathfrak{q}_1$ of prime ideals such that $\mathfrak{q}_i \cap R = \mathfrak{p}_i$, $i = 1, \ldots, n$.*

Proof: By induction, we may assume $n = 2$, $m = 1$. Let P be the multiplicatively closed subset of S containing elements of the form rs with $r \in R - \mathfrak{p}_2$ and $s \in S - \mathfrak{q}_1$ (i.e. P" = "$(R - \mathfrak{p}_2)(S - \mathfrak{q}_1)$). Note that $R - \mathfrak{p}_2$ and $S - \mathfrak{q}_1$ are contained in P as $1 \in R, S$. Suppose we show that $S_P \mathfrak{p}_2$ is a proper ideal in S_P i.e. that $S\mathfrak{p}_2 \cap P = \phi$. Then $S_P \mathfrak{p}_2$ is contained in a prime ideal $\mathfrak{m}$ of S_P (e.g. a maximal ideal). But $\mathfrak{q}_2 = \mathfrak{m} \cap S$ is a prime ideal and $\mathfrak{q}_2 \cap P = \phi$ so that $\mathfrak{q}_2 \subset \mathfrak{q}_1$. Moreover, $\mathfrak{q}_2 \cap R \supset \mathfrak{p}_2$. Since $\mathfrak{q}_2 \cap (R - \mathfrak{p}_2) = \phi$ (as $\mathfrak{q}_2 \cap P = \phi$), $\mathfrak{q}_2 \cap R = \mathfrak{p}_2$. So, we now show that $S\mathfrak{p}_2 \cap P = \phi$. If $u \in S\mathfrak{p}_2$, then, by lemma 16.37, u is integral over $\mathfrak{p}_2$ and, by proposition 16.38, the minimal equation of u over $K(R)$ is $f(u) = u^n + a_1 u^{n-1} + \cdots + a_n = 0$ with $a_i \in \mathfrak{p}_2$. If u were also in P, then $u = rs$ with $r \in R - \mathfrak{p}_2$, $s \in S - \mathfrak{q}_1$ so that $s = ur^{-1}$ with $r^{-1} \in K(R)$ and the minimal equation for s over $K(R)$ is

$$s^n + b_1 s^{n-1} + \cdots + b_n = 0 \qquad b_i = a_i r^{-i} \qquad (16.41)$$

where $r^i b_i = a_i \in \mathfrak{p}_2$. Since s is integral over R, the b_i are in R by proposition 16.38. If $r \notin \mathfrak{p}_2$, then each $b_i \in \mathfrak{p}_2$ and $s^n \in S\mathfrak{p}_2 \subset S\mathfrak{p}_1 \subset \mathfrak{q}_1$. But $\mathfrak{q}_1$ is prime implies the contradiction $s \in \mathfrak{q}_1$.

Corollary 16.42. *If $\mathfrak{q}$ is a minimal prime ideal in S and $\mathfrak{q} \cap R = \mathfrak{p}$, then $\mathfrak{p}$ is a minimal prime ideal in R.*

Proof: If $\mathfrak{p}_1 \subset \mathfrak{p}$, then there is a $\mathfrak{q}_1 \subset \mathfrak{q}$ with $\mathfrak{q}_1 \cap R = \mathfrak{p}_1$. But $\mathfrak{q}_1 = \mathfrak{q}$ or (0) implies $\mathfrak{p}_1 = \mathfrak{p}$ or (0).

We shall use the corollary to prove a converse of corollary 16.13 and then, prove the key result relating the dimension of a prime ideal $\mathfrak{p}$ to its depth. We first need a lemma known as the Noether Normalization lemma.

Lemma 16.43. *Let $R = k[u_1, \ldots, u_N]$ be a finitely generated in-*

tegral domain with transcendence degree ν over k. Then there are algebraically independent elements $z_1, \ldots, z_\nu$ of R such that R is integral over $R' = k[z_1, \ldots, z_\nu]$.

Proof: If $\nu = N$, then take $z_i = u_i$ and so we use induction on N. Suppose $N > \nu$ and that the result is true for $N - 1$ generators. Since $N > \nu$, $u_1, \ldots, u_N$ are algebraically dependent over k. Let $f \neq 0$ be a polynomial with $f(u_1, \ldots, u_N) = 0$ and suppose that f is of degree d. Set

$$v_i = u_i - u_1^{(d+1)^{i-1}} \tag{16.44}$$

for $i = 2, \ldots, N$. Then $f(u_1, v_2 + u_1^{(d+1)}, \ldots, v_N + u_1^{(d+1)^{N-1}}) = 0$ i.e. $u_1, v_2, \ldots, v_N$ are roots of the polynomial $f(X_1, Y_2 + X_1^{(d+1)}, \ldots, Y_N + X_1^{(d+1)^{N-1}})$. Each term of f gives rise to a monomial of the form $\alpha_e X_1^e$ where $e = e_1 + e_2(d+1) + \cdots + e_N(d+1)^{N-1}$. As the e_i's range from 0 to d, the resulting e's are all distinct and one of them will be the term of highest order. In other words, for some e, $f(X_1, Y_2 + X_1^{(d+1)}, \ldots, Y_N + X_1^{(d+1)^{N-1}}) = \alpha_e X_1^e +$ terms of degree $< e$ with $\alpha_e \neq 0$. It follows that $\alpha_e^{-1} f(u_1, v_2 + u_1^{(d+1)}, \ldots,) = 0$ is an equation of integral dependence for u_1 over $k[v_2, \ldots, v_N]$ and hence, R is integral over $k[v_2, \ldots, v_N]$. The lemma follows by induction and the transitivity of integral dependence.

Theorem 16.45. *If $\mathfrak{p}$ is a minimal prime ideal in the finitely generated integral domain $R = k[x_1, \ldots, x_N]$ of transcendence degree ν, then $\dim_k \mathfrak{p} = \nu - 1$.*

Proof: If $\nu = N$, then R is a polynomial ring and hence a unique factorization domain. Therefore, $\mathfrak{p} = (f)$ with f prime and degree $f > 0$. Say x_N occurs in f, then $g(\overline{x}_1, \ldots, \overline{x}_{N-1}) = 0$ ($\overline{x}_i$ in $R/\mathfrak{p}$) would imply that f divides g and x_N occurs in g. This is a contradiction. Therefore, tr. $\deg(R/\mathfrak{p})/k = \dim_k \mathfrak{p} \geq \nu - 1$. But $\dim_k \mathfrak{p} \leq \nu - 1$ by corollary 16.11. If $\nu < N$, we use the Noether Normalization lemma to give R integral over $R' = k[z_1, \ldots, z_\nu]$. Let $\mathfrak{p}' = \mathfrak{p} \cap R'$. Since R' is a polynomial ring, R' is integrally closed. But $\mathfrak{p}'$ is minimal as $\mathfrak{p}$ is (corollary 16.42) and, by what was just proved, $\dim_k \mathfrak{p}' = \nu - 1$. Since $R/\mathfrak{p}$ is integral over $R'/\mathfrak{p}'$ (proposition 16.36), $\dim_k \mathfrak{p} = \nu - 1$.

Theorem 16.46. *Let R be a finitely generated integral domain of transcendence degree ν over k and let $\mathfrak{p}$ be a prime ideal in R. If $\dim_k \mathfrak{p} = s$, then*

$$h(\mathfrak{p}) = \nu - s \; , \quad d(\mathfrak{p}) = s \; , \quad h(\mathfrak{p}) + d(\mathfrak{p}) = \nu \qquad (16.47)$$

so that $\dim_k \mathfrak{p} = d(\mathfrak{p})$.

Proof: For $d(\mathfrak{p})$, we use induction from $s - 1$ to s. We may assume $s \neq 0$ (otherwise $\mathfrak{p}$ is maximal and corollary 16.12 applies). If $\mathfrak{p} = \mathfrak{p}_d < \cdots < \mathfrak{p}_0 < R$, then $0 = \dim_k \mathfrak{p}_0 < \cdots < \dim_k \mathfrak{p} = s$ and so, $d(\mathfrak{p}) \leq s$. Consider $R/\mathfrak{p}$ and let $\overline{\mathfrak{p}}'$ be a minimal prime ideal in $R/\mathfrak{p}$ where $\mathfrak{p}' > \mathfrak{p}$ is a prime ideal with $\mathfrak{p}'/\mathfrak{p} = \overline{\mathfrak{p}}'$. Then $\dim_k \overline{\mathfrak{p}}' = s - 1 = \dim_k \mathfrak{p}'$. But then $d(\mathfrak{p}') = s - 1$ and $d(\mathfrak{p}) \geq d(\mathfrak{p}') + 1 = s$.

For $h(\mathfrak{p})$, we use induction from $s + 1$ to s and we may assume $\mathfrak{p} \neq (0)$. If $(0) < \mathfrak{p}_1 < \cdots < \mathfrak{p}_h = \mathfrak{p}$, then $\nu = \dim_k(0) > \dim_k \mathfrak{p}_1 > \cdots > \dim_k \mathfrak{p} = s$ and so, $h(\mathfrak{p}) \leq \nu - s$. Since $\mathfrak{p} \neq (0)$, there is a prime ideal $\mathfrak{p}' < \mathfrak{p}$ such that there are no prime ideals between $\mathfrak{p}'$ and $\mathfrak{p}$. In other words, $\mathfrak{p}/\mathfrak{p}'$ is a minimal prime ideal in the finite integral domain $R/\mathfrak{p}'$. Therefore, $\dim_k \mathfrak{p}/\mathfrak{p}' = \dim_k \mathfrak{p}' - 1$. But $\dim_k \mathfrak{p}/\mathfrak{p}' = \text{tr. deg}(R/\mathfrak{p}')/(\mathfrak{p}/\mathfrak{p}') = \text{tr. deg } R/\mathfrak{p}$ (as $(R/\mathfrak{p}')/(\mathfrak{p}/\mathfrak{p}') = R/\mathfrak{p}) = \dim_k \mathfrak{p}$. Hence, $\dim_k \mathfrak{p}' = s + 1$ and $h(\mathfrak{p}') = \nu - (s + 1)$. But $h(\mathfrak{p}') = h(\mathfrak{p}) - 1$ and so, $h(\mathfrak{p}) = \nu - s$.

Corollary 16.48. *Let V be an affine variety. Then*

$$\text{Dim}_k V = \dim_k I(V) = \dim_k V \qquad (16.49)$$

i.e. the topological and algebraic dimensions of V over k are the same.

Proof: Closed irreducible subsets of V correspond to prime ideals in $k[X_1, \ldots, X_N]$ which contain $I(V)$ and hence to prime ideals in $k[V]$. Thus, $\dim_k V = d(I(V)) = \dim_k I(V) = \text{tr. deg } k[V] = \text{Dim}_k V$.

Corollary 16.50. *If V is a quasi-affine variety, then $\dim_k V = \dim_k \overline{V}$ where $\overline{V}$ is the (Zariski) closure of V.*

Proof: If $V_0 < V_1 < \cdots < V_\mu$ is a chain of irreducibles in V, then $\overline{V}_0 < \overline{V}_1 < \cdots < \overline{V}_\mu$ is a chain of irreducibles in $\overline{V}$ by proposition 6.3

and so $\dim_k V \leq \dim_k \overline{V}$. Let $\nu = \dim_k V$ and let $V_0 < \cdots < V_\nu$ be a maximal chain in V so that $V_0 = \{v_0\}$ is a point. Then, again by proposition 6.3, $\{v_0\} < \overline{V}_1 < \cdots < \overline{V}_\nu$ is a maximal chain in $\overline{V}$. Since $\mathfrak{m}_0 = I(\{v_0\}) \cap k[\overline{V}]$ is a maximal ideal in $k[\overline{V}]$ and the $\overline{V}_i$ correspond to prime ideals in $k[\overline{V}]$, we have $h(\mathfrak{m}_0) = \nu$. Since $\mathfrak{m}_0$ is maximal, $d(\mathfrak{m}_0) = 0$. It follows from the theorem that $h(\mathfrak{m}_0) + d(\mathfrak{m}_0) = \nu = \dim_k \overline{V}$.

Corollary 16.51. $\dim_k GL(n,k) = n^2$, $\dim_k \mathrm{Rat}(n,k) = 2n$, $\dim_k \mathrm{Hank}(n,k) = 2n$.

Corollary 16.52. *An affine variety $V \subset \mathbb{A}_k^N$ has dimension $N - 1$ (or, in other words, codimension 1) if and only if $V = V(f)$ where f is a prime in $k[X_1, \ldots, X_N]$.*

Proof: The "if" has been established in corollary 16.13. On the other hand, if f is a prime, then (f) is a minimal prime ideal and so, $\dim_k V(f) = \dim_k (f) = N - 1$.

From now on, we shall speak of the *dimension of V* and use the notation, $\dim V$ (or $\dim_k V$). We also note that $\dim V$ is invariant under isomorphism (corollary 16.8) and that if $V_1 < V_2$ with V_2 irreducible, then $\dim V_1 < \dim V_2$ (corollary 16.10). Since $o_G(x_0)$ and $G = GL(n,k)$ are k-isomorphic for any x_0 in $S_{1,1}^n$, we have, in view of the "Closed Orbit Lemma" (lemma 15.9), the following:

Proposition 16.53. *The orbits $o_G(x_0)$, $x_0 \in S_{1,1}^n$, are closed.*

Thus, again, we have established the first property of a geometric quotient for $(\mathrm{Char}(n,k), \mathfrak{R}_\chi)$.

Now let us reflect for a moment on corollary 16.52 and corollaries 5.15 and 5.16. The first states that a variety in $\mathbb{A}_k^N$ defined by a single equation has dimension $N - 1$ and the latter states that a variety defined by N equations has dimension 0. Suppose we have a variety defined by r equations or that $\mathbb{A}_k^N$ is replaced by an irreducible algebraic set of dimension N. What can we say under these circumstances? We begin with the following geometric version of Krull's Principal Ideal Theorem ([M-2]).

Theorem 16.54. *Let V be an affine variety, $0 \neq f \in k[V]$ with f a* *nonunit in $k[V]$, and let Z be an irreducible component of $V(f) = \{v \in V : f(v) = 0\}$. Then $\dim Z = \dim V - 1$ so that the codimension of Z in V is 1.*

Proof: Let $\mathfrak{p} = I(Z)$ in $k[V]$ and let Z_i, $i = 1, \ldots, s$ be the other irreducible components of $V(f)$ with $\mathfrak{p}_i = I(Z_i)$. Then $\sqrt{(f)} = \mathfrak{p} \cap \mathfrak{p}_1 \cap \cdots \cap \mathfrak{p}_s$ (by the Nullstellensatz). Since the decomposition into irreducible components contains no superfluous elements, there is a $g \in \cap_{i=1}^{s} \mathfrak{p}_i$, $g \notin \mathfrak{p}$. Let us replace V by V_g and $k[V]$ by $k[V_g] = k[V]_g$. Then $V_g(f) = \{v \in V_g : f(v) = 0\} = Z \cap V_g$ is irreducible and $I(Z \cap V_g)$ in $k[V]_g$ is simply $\mathfrak{p}k[V]_g$. In other words, we may assume that $Z = V(f)$ and that $\mathfrak{p} = I(Z) = \sqrt{(f)}$. Let $S = k[V]$. Then by the Noether Normalization Lemma (16.43), there are algebraically independent elements $z_1, \ldots, z_\nu$ of S, $\nu = \dim V = \text{tr. deg } S/k$, such that S is integral over $R = k[z_1, \ldots, z_\nu]$. Since R is integrally closed (corollary 16.29) and $K(S) = k(V)$ is a finite algebraic extension of $K(R)$ (as R is finitely generated over k), we deduce immediately from corollary 16.39 that $\mathfrak{p} \cap R = \sqrt{(f_0)}$ where $f_0 = N_{K(S)/K(R)}(f)$. Since R is a unique factorization domain, $f_0 = \alpha f_0'^t$ where α is a unit and f_0' is a prime in R. In other words, $\mathfrak{p} \cap R = (f_0')$. It follows from corollary 16.52 that $\text{tr. deg}_k R/R \cap \mathfrak{p} = \nu - 1$. Since $S/\mathfrak{p}$ is integral over $R/R \cap \mathfrak{p}$ (proposition 16.36), we have $\dim Z = \text{tr. deg}_k S/\mathfrak{p} = \text{tr. deg}_k R/R \cap \mathfrak{p} = \nu - 1 = \dim V - 1$.

The theorem has some significant corollaries.

Corollary 16.55. *Let V be an affine variety and let Z be a closed irreducible subvariety of V with codim $Z = 1$. Then if U is open in V with $Z \cap U \neq \phi$ and if $f \in k[U]$, $f \neq 0$ with $f(Z) = 0$, then $Z \cap U$ is an irreducible component of $V(f)$.*

Proof: If W is the irreducible component of $V(f)$ with $Z \cap U \subset W$, then $\dim U > \dim W \geq \dim(Z \cap U) = \dim U - 1 = \dim V - 1$ and so, $W = Z \cap U$ by corollary 16.10.

Corollary 16.56. *Let V be an affine variety and let W be a closed irreducible subset of codimension $r \geq 1$. Then there exists a chain of closed irreducible subsets Z_i, $i = 1, \ldots, r$ of V such that $Z_1 > Z_2 >$*

$\ldots > Z_r = W$ *and* $\operatorname{codim} Z_i = i$.

Proof: Use induction on r. For $r = 1$, $Z_r = W$. Since $W < V$, there is an $f \neq 0$ ($f \in k[V]$) in $I(W)$ such that $W \subset Z_1$ where Z_1 is an irreducible component of $V(f)$. By the theorem, $\operatorname{codim} Z_1 = 1$ and by induction, there are Z_i, $i = 2, \ldots, r$ such that $Z_2 > \cdots > Z_r = W$ and $\operatorname{codim} Z_i = i$.

Note that this corollary also shows that the topological and algebraic notions of dimension are the same.

Corollary 16.57. *Let* $f_1, \ldots, f_r \in k[V]$ *and let* Z *be an irreducible component of* $V(f_1, \ldots, f_r)$. *Then* $\operatorname{codim} Z \leq r$.

Proof: Use induction on r. For $r = 1$, the theorem applies. Since Z in an irreducible subset of $V(f_1, \ldots, f_{r-1})$, $Z \subset Z_1$ where Z_1 is a component of $V(f_1, \ldots, f_{r-1})$. Since Z is a maximal irreducible subset of $V(f_1, \ldots, f_r)$, Z is a component of $Z_1 \cap V(f_r)$. By induction $\operatorname{codim} Z_1 \leq r - 1$. If $f_r(Z_1) = 0$, then $Z = Z_1$ and $\operatorname{codim} Z \leq r - 1 \leq r$. If $f_r \neq 0$ on Z_1, then $\dim Z = \dim Z_1 - 1$ by the theorem and $\operatorname{codim} Z = \dim V - \dim Z = \dim V - \dim Z_1 + 1 = \operatorname{codim} Z_1 + 1 \leq r$.

Corollary 16.58. *Let* V *be an affine variety and let* Z *be a closed irreducible subset of codimension* $r \geq 1$. *Then there are* $f_1, \ldots, f_r$ *in* $k[V]$ *such that* Z *is a component of* $V(f_1, \ldots, f_r)$.

Proof: We shall actually prove a more general result. Let $Z_1 > \cdots > Z_r = Z$ be a chain with $\operatorname{codim} Z_i = i$ (corollary 16.56). Then there are $f_1, \ldots, f_r$ in $k[V]$ such that Z_i is a component of $V(f_1, \ldots, f_i)$ and *all* components of $V(f_1, \ldots, f_i)$ have codimension i.

We use induction on i. For $i = 1$, let $f_1 \in I(Z_1)$ with $f_1 \neq 0$ and apply corollary 16.54. Suppose then, that $f_1, \ldots, f_{i-1}$ have been determined and that $Z_{i-1} = W_1, W_2, \ldots, W_s$ are the components of $V(f_1, \ldots, f_{i-1})$. Since each W_j has codimension $i - 1$, none is contained in Z_i. Therefore, $I(W_j) \not\supset I(Z_i)$ for $j = 1, \ldots s$. Since the $I(W_j)$ are prime, $\cup_{j=1}^{s} I(W_j) \not\supset I(Z_i)$ (exercise 48). Let $f_i \in I(Z_i)$ such that $f_i \notin \cup_{j=1}^{s} I(W_j)$. If W is any component of $V(f_1, \ldots, f_i)$, then W is a component of $W_j \cap V(f_i)$ for some j (as in the proof of corollary 16.57). Since $f_i \neq 0$ on W_j, $\dim W = \dim W_j - 1$ and codim $W = i$. But, $Z_i \subset V(f_1, \ldots, f_i)$ as $f_i \in I(Z_i)$. Since Z_i is irreducible, Z_i is contained in a component W of $V(f_1, \ldots, f_i)$. But codim $Z_i = i = $ codim W and so, $Z_i = W$.

We shall use these corollaries in studying fibers of morphisms.

17. The Geometric Quotient Theorem: Open On Invariant Sets

We now turn our attention to the second property of a geometric quotient. In other words, we want to prove the following:

Theorem 17.1. *If U is an invariant open set in $S_{1,1}^n$, then there is an open set U_0 in $\mathrm{Char}(n, k)$ such that $\psi^{-1}(U_0) = U$ (where $\psi = \mathfrak{R}_\chi$ as in section 15.)*

As noted earlier, we shall give several proofs of this theorem.

Definition 17.2. A morphism $\varphi : V \to W$ of varieties is *dominant* if $\varphi(V)$ is dense in W i.e. $\overline{\varphi(V)} = W$.

We observe that, in view of proposition 7.18, φ is dominant if and only if φ^* is injective. Since the mapping $\psi : S_{1,1}^n \to \mathrm{Char}(n, k)$ is surjective, ψ is a dominant morphism. Furthermore, we have:

Proposition 17.3. *If $\xi \in \mathrm{Char}(n, k)$, then the "fiber" $\psi^{-1}(\xi) = \{x \in S_{1,1}^n : \psi(x) = \xi\}$ is a closed irreducible variety and $\dim_k \psi^{-1}(\xi) = n^2$. Hence, all components of $\psi^{-1}(\xi)$ have dimension equal to $\dim_k S_{1,1}^n - \dim_k \mathrm{Char}(n, k)$.*

Proof: Simply note that $\psi^{-1}(\xi) = o_G(x)$ for some x in $S_{1,1}^n$ and that $o_G(x)$ and $G = GL(n, k)$ are k-isomorphic by the State Space Isomorphism theorem (11.19).

Let us, for the moment, call an affine (or quasi-affine) variety V *normal* if its coordinate ring $k[V]$ is integrally closed. Then $\mathrm{Char}(n, k)$ is normal by virtue of corollary 16.35. Now, summarizing, we observe that $\psi : S_{1,1}^n \to \mathrm{Char}(n, k)$ is a *dominant morphism* into a *normal* variety such that *each fiber* $\psi^{-1}(\xi)$ has *all* irreducible *components of dimension* $n^2 = \dim_k S_{1,1}^n - \dim_k \mathrm{Char}(n, k)$. It follows from proposition 18.22 of the next section that ψ *is an open map*. We now have:

Proof 1. (of Theorem 17.1).   Suppose then that ψ is an open map and let U be an open invariant set in $S_{1,1}^n$. Then $U_0 = \psi(U)$ is open in $\text{Char}(n,k)$ and $\psi^{-1}(U_0)$ is an open invariant set in $S_{1,1}^n$ with $U \subset \psi^{-1}(U_0)$. We claim that $U = \psi^{-1}(U_0)$. Let $x_0 \in \psi^{-1}(U_0)$ so that $\psi(x_0) = \xi_0 \in U_0 = \psi(U)$. Then there is an x in U with $\psi(x) = \xi_0$ and hence, $x_0 \in o_G(x)$ (by Theorem 11.19). But U is invariant means that $o_G(x) \subset U$ and hence that $x_0 \in U$.

Of course, the difficult part of this proof is embodied in proposition 18.22 which shall be proved in the next section. Now, in view of the results in Appendix B (and [F-2], [M-3], [M-4]), we note that the equivalence relation E_G, $G = GL(n,k)$, generated by the action

$$g \cdot (X,Y,Z) = (gXg^{-1}, gY, Zg^{-1}) \tag{17.4}$$

on $R_\theta = k[(X_{ij}), (Y_j), (Z_i)]_\theta = k[X,Y,Z]_\theta$ where $\theta(X,Y,Z) = \det \mathbf{H}(X,Y,Z)$, is *proper*. We then have:

Proof 2. (of Theorem 17.1).   Simply apply corollary 14.16.

Of course, again the difficult part of this proof is embodied in the results in Appendix B and in [F-2], [M-3] and [M-4].

18. Affine Algebraic Geometry: Fibers of Morphisms

We shall in this section use the term, "variety", to mean either an affine or quasi-affine variety. Our goal is to study the fibers of morphisms and, in particular, to examine the dimensions of fibers. We begin with

Definition 18.1. Let $\psi : X \to Y$ be a morphism of varieties. If $y \in Y$, the closed set $\psi^{-1}(y) = \{x \in X : \psi(x) = y\}$ is called the *fiber of ψ over y*. If W is a closed irreducible subset of Y, then the closed set $\psi^{-1}(W) = \{x \in X : \psi(x) \in W\}$ is called the *fiber of ψ over W*.

Example 18.2. Let $L : \mathbb{A}_k^2 \to \mathbb{A}_k^2$ be given by $L(x, y) = (x, -xy)$. Then $L^{-1}((0,0)) = \{(0,y), y \in k\}$ and so $\dim L^{-1}((0,0)) = 1 > \dim \mathbb{A}_k^2 - \dim \mathbb{A}_k^2$.

Example 18.3. Let $\psi : \mathbb{A}_k^1 \to \mathbb{A}_k^2$ be given by $\psi(t) = (t^2 - 1, t(t^2 - 1))$. The image of ψ is the curve $Y^2 = X^2(X + 1)$ and $(0,0)$ is a point on this curve. But $\psi^{-1}((0,0)) = \{1\} \cup \{-1\}$ and so a fiber may well be reducible.

Definition 18.4. A morphism $\psi : X \to Y$ of varieties is *dominant* if $\overline{\psi(X)} = Y$ i.e. $\psi(X)$ is dense in Y. If Z is a component of $\psi^{-1}(W)$, then Z *dominates* W if $\psi(Z)$ is dense in W.*

We observe that if $\psi : X \to Y$ is any morphism of varieties, then $Z = \overline{\psi(X)}$ is irreducible and the morphism $\overline{\psi} : X \to Z$ given by $\overline{\psi}(x) = \psi(x)$ is dominant. If $i : Z \to Y$ is the natural injection, then $\psi = i \circ \overline{\psi}$ and so, the study of fibers of morphisms is essentially reduced to the case of dominant morphisms.

---

* When X, Y are not irreducible, we use the term *dominant* to mean both $\overline{\psi(X)} = Y$ and ψ maps each component of X onto a dense subset of a component of Y.

Theorem 18.5. *Let $\psi : X \to Y$ be a dominant morphism of varieties and let W be a closed irreducible subset of Y. If Z is a component of $\psi^{-1}(W)$ which dominates W, then $\dim Z \geq \dim W + \dim X - \dim Y$.*

Proof: Since ψ is dominant, ψ^* is an injection of $k[Y]$ into $k[X]$ so that $\dim X \geq \dim Y$ (proposition 7.18). If $r = $ codimension of W in Y, then W is a component of $V(g_1, \ldots, g_r)$ for $g_i \in k[Y]$ by corollary 16.58. Let $f_i = \psi^*(g_i)$. Then $Z \subset \psi^{-1}(W) \subset V(f_1, \ldots, f_r)$. Since Z is irreducible, $Z \subset Z'$ where Z' is a component of $V(f_1, \ldots, f_r)$. But $W = \overline{\psi(Z)} \subset \overline{\psi(Z')} \subset V(g_1, \ldots, g_r)$. Since W is a component of $V(g_1, \ldots, g_r)$, $W = \overline{\psi(Z')}$ and $Z' \subset \psi^{-1}(W)$. But Z is a component of $\psi^{-1}(W)$ and so $Z = Z'$ is a component of $V(f_1, \ldots, f_r)$. It follows from corollary 16.57 that $\text{codim } Z \leq r = \text{codim } W$. In other words, $\dim X - \dim Z \leq \dim Y - \dim W$.

Corollary 18.6. $\dim \psi^{-1}(y) \geq \dim X - \dim Y$, *for $y \in Y$.*

We now make the following:

Definition 18.7.   A morphism $\psi : X \to Y$ of varieties is *finite* if $k[X]$ is integral over $\psi^*(k[Y])$.

If ψ is a dominant finite morphism then $k(X)$ is a finite algebraic extension of $\psi^*(k(Y))$ and so, $\dim X - \dim Y = 0$. The term finite comes from the fact that the fibers $\psi^{-1}(y)$ of a finite dominant morphism are finite sets.

Example 18.8.   (cf. example 7.23) Let $X = \mathbb{A}^1_k$ and let $Y \subset \mathbb{A}^2_k$ be given by $Y = \{(\xi_1, \xi_2) : \xi_1^3 - \xi_2^2 = 0\}$. The morphism $\psi : X \to Y$ given by $\psi(\xi) = (\xi^2, \xi^3)$ is finite and dominant. This is so since $k[Y] \simeq k[t^2, t^3]$ and $k[X] = k[t]$ is integral over $k[Y]$.

Before developing the properties of fibers further, we shall need some results on integral extensions. In particular, we prove the so-called "Going Up" theorem. We begin with some lemmas.

Lemma 18.9. *Let R be a subring of S and suppose that S is integral over R. If S is a field, then R is a field.*

Proof: Let $r \neq 0$ be an element of R. Then $1/r \in S$ as S is a field and $1/r$ is integral over R. Thus, $(1/r)^n + r_1(1/r)^{n-1} + \cdots + r_n = 0$ with $r_i \in R$. It follows that $1/r = -r_1 - \cdots - r_n r^{n-1} \in R$.

Lemma 18.10. *Let R be a local ring with $\mathfrak{m}$ as unique maximal ideal. Suppose that R is a subring of S and that S in integral over R. If $\mathfrak{m}'$ is a maximal ideal of S, then $\mathfrak{m}' \cap R = \mathfrak{m}$.*

Proof: We note that $\mathfrak{m}' \cap R \subset \mathfrak{m}$ and that $R/\mathfrak{m}' \cap R$ may naturally be viewed as a subring of $S/\mathfrak{m}'$. Moreover, $S/\mathfrak{m}'$ is integral over $R/\mathfrak{m}' \cap R$. Since $S/\mathfrak{m}'$ is a field, $R/\mathfrak{m}' \cap R$ is a field by the previous lemma. Hence, $\mathfrak{m}' \cap R$ is a maximal ideal and so, $\mathfrak{m}' \cap R = \mathfrak{m}$.

Theorem 18.11 (Going Up). *Let R be a subring of S and suppose that S is integral over R. If $\mathfrak{p}$ is a prime ideal of R, then there exist prime ideals $\mathfrak{p}'$ of S such that $\mathfrak{p}' \cap R = \mathfrak{p}$.*

Proof ([M-2]). We reduce to the case of lemma 18.10 by localizing both R and S with respect to the multiplicative set $M = R - \mathfrak{p}$. In other words, $R_{\mathfrak{p}} = R_M$ is a local ring which is a subring of S_M and S_M is integral over R_M. Moreover, if i_R, i_S denote the natural maps of R into R_M and S into S_M, then $\mathfrak{p} = i_R^{-1}(\mathfrak{p}R_{\mathfrak{p}})$ and if $\tilde{\mathfrak{p}}$ is a prime ideal of S_M such that $\tilde{\mathfrak{p}} \cap R_M = \mathfrak{p}R_{\mathfrak{p}}$, then $\mathfrak{p}' = i_S^{-1}(\tilde{\mathfrak{p}})$ is a prime ideal in S with $\mathfrak{p}' \cap R = i_R^{-1}(\tilde{\mathfrak{p}} \cap R_M) = i_R^{-1}(\mathfrak{p}R_{\mathfrak{p}}) = \mathfrak{p}$.

We shall now prove a critical theorem on finite morphisms.

Theorem 18.12. *Let $\psi : X \to Y$ be a dominant finite morphism of varieties. Then (a) ψ is closed (and therefore surjective); (b) if W is a closed irreducible subset of Y and Z is a component of $\psi^{-1}(W)$, then $\psi(Z)$ is closed, $\dim\psi(Z) = \dim Z$ and, for some Z, $\psi(Z) = W$; and, (c) if $k[Y]$ is integrally closed (i.e. Y is normal), then $\psi(Z) = W$ for all components Z of $\psi^{-1}(W)$.*

Proof ([D-1], [H-7]).

(a) Let $R = k[Y]$ and $S = k[X]$. Since ψ^* is injective, we shall view R as a subring of S with S integral over R. If $\mathfrak{a}$ is an ideal in S, then $S/\mathfrak{a}$ is naturally an integral extension of $R/\mathfrak{a} \cap R$. If $Z = V(\mathfrak{a})$ is closed in X where $\mathfrak{a} = I(Z) = \sqrt{\mathfrak{a}}$, then $\psi(Z) \subset V(\mathfrak{a} \cap R) = Z'$ and $I(Z') = \mathfrak{a} \cap R$ (since $\mathfrak{a} = \sqrt{\mathfrak{a}}$). But $k[Z] = k[X]/I(Z)$

and $k[Z'] = k[Y]/I(Z')$. It follows that $k[Z] = S/\mathfrak{a}$ and that $k[Z'] = R/\mathfrak{a} \cap R$. Hence, $\bar{\psi} : Z \to Z'$ ($\bar{\psi}$ is the restriction of ψ) is again finite and dominant. Thus, it is enough to show the following:

(18.13) any finite dominant morphism is surjective.

But if $y \in Y$ and $\mathfrak{m}_y$ is the maximal ideal of y in R, then, by the Going Up theorem, there is a maximal ideal $\mathfrak{m}_x$ in S such that $\mathfrak{m}_x \cap R = \mathfrak{m}_y$. Thus, $\psi(x) = y$ and ψ is surjective.

(b) Let Z be a component of $\psi^{-1}(W)$. Then, by part (a), $\psi(Z)$ is closed. Morever, if $\bar{\psi}$ is the restriction of ψ to Z, then $\bar{\psi} : Z \to \psi(Z)$ is finite and dominant so that $\dim Z = \dim \psi(Z)$. Now let $\mathfrak{p} = I(W)$ in $k[Y] = R$ and let $\mathfrak{q} = I(\psi^{-1}(W))$ in $k[X] = S$. Then $\mathfrak{p}$ is prime and $\mathfrak{q} \cap R = \mathfrak{p}$. Let $\mathfrak{q} = \mathfrak{q}_1 \cap \cdots \cap \mathfrak{q}_r$ where $\mathfrak{q}_i = I(Z_i)$ and the Z_i are the components of $\psi^{-1}(W)$. Then $\mathfrak{p} = \cap_{i=1}^r (\mathfrak{q}_i \cap R)$ and, since $\mathfrak{p}$ is prime, $\mathfrak{p} = \mathfrak{q}_i \cap R$ for some i. We assert that $\psi(Z_i) = W$. For if $g \in I(\psi(Z_i))$, then $\psi^*(g) \in I(Z_i) = \mathfrak{q}_i$ and so, $\psi^*(g) \in \mathfrak{q}_i \cap R = \mathfrak{p}$. Since ψ^* is injective (actually here $\psi^*(g) = g$), $g \in I(W)$ i.e. $I(\psi(Z_i)) \subset I(W)$. It follows that $\psi(Z_i) = W$.

(c) Let $\mathfrak{p}$, $\mathfrak{q}$, $\mathfrak{q}_i$ be as in part (b). Then $\mathfrak{p} = \mathfrak{q} \cap R \subset \mathfrak{q}_i \cap R$ for all i. By the Going Down Theorem, there is, for each i, a prime ideal $\mathfrak{q}_i' \subset \mathfrak{q}_i$ such that $\mathfrak{q}_i' \cap R = \mathfrak{p}$. But $\psi(V(\mathfrak{q}_i')) \subset W$ and so, $V(\mathfrak{q}_i')$ is a closed irreducible subset of $\psi^{-1}(W)$. Since $Z_i = V(\mathfrak{q}_i) \subset V(\mathfrak{q}_i')$, we must have $Z_i = V(\mathfrak{q}_i')$ and $\mathfrak{q}_i' = \mathfrak{q}_i$. In other words, $\mathfrak{q}_i \cap R = \mathfrak{p}$ for each i. By the proof of part (b), it follows that $\psi(Z_i) = W$ for all i.

Corollary 18.14. *A morphism ψ is finite and dominant if and only if ψ is finite and surjective*

Corollary 18.15. *If $\psi : X \to Y$ is a finite surjective morphism, if Y is normal i.e. $k[Y]$ is integrally closed, and W is a closed irreducible subset of Y, then $\dim Z = \dim W = \dim W + \dim X - \dim Y$ for every component Z of $\psi^{-1}(W)$.*

Now, theorem 18.5 says that the dimension of the fibers is not "too small". Example 18.2 indicates that the dimension of the fibers of a morphism may not be "just right" (i.e. equal to $\dim X - \dim Y$)

everywhere. We shall soon show that things work out "almost every-where".

Lemma 18.16. *Let $R \subset S$ be finitely generated integral domains over k. Then there are $x_1, \ldots, x_\nu$ in S and f in R such that S_f is integral over $R_f[x_1, \ldots, x_\nu]$, $\nu = \mathrm{tr.\ deg}\, K(S)/K(R)$.*

Proof: (Exercise 49)

Theorem 18.17. *Let $\psi : X \to Y$ be a dominant morphism of varieties. Then there is an open set $U \subset Y$ such that (a) $U \subset \psi(X)$ and (b) if W is a closed irreducible subset of Y with $W \cap U \neq \phi$ and if Z is a component of $\psi^{-1}(W)$ with $Z \cap \psi^{-1}(U) \neq \phi$, then $\dim Z = \dim W + \dim X - \dim Y$.*

Proof ([H-7]).   Let $R = k[Y]$ and $S = k[X]$. View R as a subring of S. By lemma 18.16, there are $x_1, \ldots, x_\nu$ in S and f in R such that S_f is integral over $R_f[x_1, \ldots, x_\nu]$ where $\nu = \mathrm{tr.\ deg}\, k(X)/k(Y) = \dim X - \dim Y$. But $S_f = k[X_f]$ and $R_f = k[Y_f]$. Thus, $k[Y_f \times \mathbb{A}_k^\nu] = R_f[x_1, \ldots, x_\nu]$. Let $\overline{\psi}$ denote the restriction of ψ to X_f. Then $\overline{\psi} = \pi_1 \circ \varphi$ where $\varphi : X_f \to Y_f \times \mathbb{A}_k^\nu$ and $\pi_1 : Y_f \times \mathbb{A}_k^\nu \to Y_f$ is the projection. But φ (given by the injection of $R_f[x_1, \ldots, x_\nu]$ into S_f) is finite and dominant and so is surjective. Since π_1 is also surjective, $\overline{\psi}$ is surjective. Thus, if $U = Y_f$, then $\psi^{-1}(U) = X_f$ and $U \subset \psi(X)$. This establishes part (a). As for part (b), we may suppose already that $X = X_f$, $U = Y = Y_f$ and $\psi = \pi_1 \circ \varphi$. If W is a closed irreducible subset of Y and Z is a component of $\psi^{-1}(W)$, then Z is a component of $\varphi^{-1}(W \times \mathbb{A}_k^\nu)$. Since φ is finite and dominant, $\varphi(Z) = W \times \mathbb{A}_k^\nu$ and $\dim Z = \dim \varphi(Z) = \dim W + \nu = \dim W + \dim X - \dim Y$.

The process of factoring a morphism through a finite morphism (which is essentially a geometric form of normalization lemma) is often useful. We shall use the same idea in proposition 18.22. Now we make the following:

Definition 18.18.   A subset of a topological space is *locally closed* if it is the intersection of an open set and a closed set. A subset is *constructible* if it is a finite union of locally closed subsets.

The following theorem of Chevalley is often quite useful.

Theorem 18.19 (Chevalley). *Let* $\psi : X \to Y$ *be a morphism. Then* ψ *maps constructible sets into constructible sets and, in particular,* $\psi(X)$ *is constructible.*

Proof: Since a locally closed subset of an affine or quasi-affine algebraic set is again an affine or quasi-affine algebraic set, it is enough to prove that $\psi(X)$ is constructible. In addition, we may assume that $\psi : X \to Y$ is a morphism of varieties. We use induction on $\dim Y$ as the case $\dim Y = 0$ is trivial. If ψ is not dominant, then $\psi(X) \subset \overline{\psi(X)} < Y$ and $\dim \overline{\psi(X)} < \dim Y$ so the induction hypothesis applies. Thus, we suppose that ψ is dominant. If $U \subset Y$ is the non empty open set of theorem 18.17 and $W_1, \ldots, W_r$ are the components of $Y - U$ and Z_{ij} are the components of $\psi^{-1}(W_i)$, then $\psi(Z_{ij})$ is constructible since $\dim W_i < \dim Y$. Hence, $\psi(X) = U \cup \bigcup_{i,j} \psi(Z_{ij})$ is also constructible.

Corollary 18.20. *Let* $\psi : X \to Y$ *be a dominant morphism of varieties. Suppose that, for every closed irreducible* $W \subset Y$ *and each component* Z *of* $\psi^{-1}(W)$, $\dim Z = \dim W + \dim X - \dim Y$. *Then* ψ *is an open map.*

Proof: If $W = \{y\}$ is a single point, then $\psi^{-1}(W)$ is not empty by the hypothesis and so ψ is surjective. Let W be a closed irreducible subset of Y and let $Z_1, \ldots, Z_r$ be the components of $\psi^{-1}(W)$. Then Z_i is also a component of $\psi^{-1}(\overline{\psi(Z_i)})$. It follows that $\dim \overline{\psi(Z_i)} + \dim X - \dim Y = \dim Z_i = \dim W + \dim X - \dim Y$ and hence, that $\dim \overline{\psi(Z_i)} = \dim W$. In other words, $\overline{\psi(Z_i)} = W$ and Z_i dominates W. If $x \in X$ and U_x is an open neighborhood of x, then we must show that $\psi(x) = y$ is in the interior of $\psi(U_x) = V_y$. If not, then $y \in \overline{Y - V_y}$ which is constructible since V_y is constructible by the thoerem. Thus, $y \in W$ where W is the closure of some locally closed set $O \cap W \subset Y - V_y$ (O is open in Y and W may be assumed irreducible so that $\overline{O \cap W} = W$). By the hypothesis, all the components Z_i of $\psi^{-1}(W)$ have the same dimension and dominate W. Since $\psi^{-1}(O)$ meets each Z_i, $\psi^{-1}(O) \cap \psi^{-1}(W)$ is dense in $\psi^{-1}(W)$. But $\psi^{-1}(O) \cap \psi^{-1}(W) = \psi^{-1}(O \cap W) \subset X - U_x$ which is closed. Then, $x \in \psi^{-1}(W) \subset X - U_x$ which is a contradiction.

Corollary 18.21. *If $\psi : X \to Y$ is a finite surjective moprhism of varieties and Y is normal (i.e. $k[Y]$ is integrally closed), then ψ is an open map.*

Proof: Apply corollary 18.15 and corollary 18.20.

We now prove the proposition which is the key to the first proof of Theorem 17.1 (the second property of a geometric quotient).

Proposition 18.22 (Chevalley [D-1]). *Let $\psi : X \to Y$ be a dominant morphism of varieties and let $r = \dim X - \dim Y$. Suppose that Y is normal and that all the components of $\psi^{-1}(y)$, $y \in Y$, are r-dimensional. Then ψ is an open map.*

Proof: We shall reduce the proposition to corollary 18.21 and proposition 12.12 by appropriately factoring ψ. Actually, we will show that if $y \in Y$ and $\psi(x) = y$ and U_x is an open neighborhood of x, then $\psi(U_x)$ is an open neighborhood of y. In fact, to do this it will be enough to show that $\psi(X)$ is an open neighborhood of $\psi(x) = y$ in Y (by restricting ψ and noting that the intersection of an open set $O \subset X$ with any component of $\psi^{-1}(y)$ is dense in that component). So let $R = k[Y]$ and $S = k[X]$ and let us view R as a subring of S. Since Y is normal, R is integrally closed. Set $\mathfrak{m}_y = I(\{y\})$. Then $\psi^{-1}(y) = V(\mathfrak{m}_y S)$ and $k[\psi^{-1}(y)] = S/\mathfrak{a}$ where $\mathfrak{a} = \sqrt{\mathfrak{m}_y S}$. Since all components of $\psi^{-1}(y)$ have dimension r, there is a k-algebra $S_1 \subset S/\mathfrak{a}$ with $S_1 = k[\overline{u}_1, \ldots, \overline{u}_r]$, the $\overline{u}_i$ algebraically independent over k, and $S/\mathfrak{a}$ integral over S_1 (by the Noether Normalization Lemma). The u_i are elements of S and the natural injection $\psi^* : R \to S$ factors as $\psi^* = \psi_1^* \circ \psi_2^*$ where $\psi_1^* : R[U_1, \ldots U_r] \to S$ is given by

$$\psi_1^*(U_i) = u_i \ , \quad i = 1, \ldots, r \qquad (18.23)$$

(the U_i being indeterminates) and $\psi_2^* : R \to R[U_1, \ldots, U_r]$ is the natural injection. In other words, $\psi = \psi_2 \circ \psi_1$ where $\psi_1 : X \to Y \times \mathbb{A}_k^r$ and $\psi_2 : Y \times \mathbb{A}_k^r \to Y$ is the projection π_Y. If $W = \psi_2^{-1}(y) = \{y\} \times \mathbb{A}_k^r$, then $\psi^{-1}(y) = \psi_1^{-1}(W)$ and, since $k[W] \simeq S_1$, the restriction of ψ_1 to $\psi^{-1}(y)$ is a finite surjective morphism and hence, an open map. Therefore, every point of $\psi^{-1}(y)$ is *isolated* in its fiber for ψ_1. But $X' = \{x' \in X : x' \text{ isolated in } \psi_1^{-1}(\psi_1(x'))\}$ is open in X and contains $\psi^{-1}(y)$. In view of theorem 18.17 and the fact that $Y \times \mathbb{A}_k^r$ is normal by the lemma which follows, ψ_1 restricted to X' is an open mapping. Since the projection ψ_2 is an open map, $\psi(X') = \psi_2(\psi_1(X'))$ is open in Y and $\psi(X)$ is a neighborhood of $\psi(x) = y$.

Lemma 18.24. *If V is a normal variety, then $V \times \mathbb{A}_k^r$ is a normal variety.*

Proof: By induction we may assume that $r = 1$. Let $R = k[V]$. Then $k[V \times \mathbb{A}_k^1] = R \otimes_k k[X] = R[X]$. Let $K = k(V)$ be the quotient field of R so that $K(X)$ is the quotient field of $R[X]$. If $f \in K(X)$ is integral over $R[X]$, then f is a fortiori integral over the unique factorizaton domain $K[X]$ and hence, $f \in K[X]$ (proposition 16.28). The result is then an immediate consequence of the following lemma.

Lemma 18.25. *Let R be an integral domain with quotient field K and let R' be the integral closure of R in K. Then $R'[X]$ is the integral closure of $R[X]$ in $K[X]$.*

Proof ([B-2]).   If $f \in K[X]$ is integral over $R[X]$, then

$$f^n + r_1 f^{n-1} + \cdots + r_n = 0 \qquad r_k \in R[X] \tag{18.26}$$

If $m > n$ and the degrees of the r_i, then $-f_1 = X^m - f$ is monic and

$$(f_1 + X^m)^n + r_1(f_1 + X^m)^{n-1} + \cdots + r_1 = 0$$

or, equivalently,

$$f_1^n + g_1 f_1^{n-1} + \cdots + g_n = 0 \tag{18.27}$$

where $g_n = (X^m)^n + r_1(X^m)^{n-1} + \cdots + r_n$. In other words, $-f_1$ and $f_1^{n-1} + g_1 f_1^{n-2} + \cdots + g_{n-1}$ are monic polynomials in $K[X]$ whose product is in $R'[X]$. Then (exercise 50.) $f_1 \in R'[X]$ and $f \in R'[X]$.

19. The Geometric Quotient Theorem: The Ring of Invariants

We now turn our attention to establishing the third property of a geometric quotient. Namely, we want to show that the ring of invariants $k[S_{1,1}^n]^G$, $G = GL(n,k)$, is $k[\text{Char}(n,k)]$. So, again as in section 15, we consider the polynomial ring $R = k[A_k^{n^2+2n}]$ which we write $k[(X_{ij}), (Y_j), (Z_i)]$, $i,j = 1, \ldots, n$ or simply $k[X,Y,Z]$. We let $\psi_i(X,Y,Z)$ be given by

$$\begin{aligned} \psi_i(X,Y,Z) &= ZX^{i-1}Y & i &= 1, \ldots, n \\ \psi_i(X,Y,Z) &= -\chi_{i-n}(X) & i &= n+1, \ldots, 2n \end{aligned} \tag{19.1}$$

where the $\chi_i's$ are the characteristic polynomials and we let $\psi : A_k^{n^2+2n} \to A_k^n$ be the regular map $(\psi_1, \ldots, \psi_{2n})$. If $R_0 = k[\text{Char}(n,k)]$, then

$$\psi^*(R_0) = k[\psi_1, \ldots, \psi_{2n}]_\theta \tag{19.2}$$

where $\theta(X,Y,Z) = \omega(X,Y,Z)\gamma(X,Y,Z) = \det \mathbf{H}(X,Y,Z)$ and $\omega(X,Y,Z) = \det \mathbf{Z}(X,Y,Z)$ and $\gamma(X,Y,Z) = \det \mathbf{Y}(X,Y,Z)$, $\mathbf{Z}(X,Y,Z)$ being the observability map and $\mathbf{Y}(X,Y,Z)$ being the controllablity map (see section 11). Then $k[S_{1,1}^n] = R_\theta$ and $k[\text{Char}(n,k)]$ " $=$ " $k[\psi_1, \ldots, \psi_n]_\theta$. We recall that $G = GL(n,k)$ acts on R via

$$g \cdot (X,Y,Z) = (gXg^{-1}, gY, Zg^{-1}) \tag{19.3}$$

and that this action preserves degrees. We also note that

Lemma 19.4. $\chi_1(X), \ldots, \chi_n(X)$, $ZX^{i-1}Y$ $i = 1, \ldots$, *are elements of $k[X,Y,Z]^G = R^G$. In other words, $k[\psi_1, \ldots, \psi_{2n}] \subset R^G$.*

Proof: Since $\det(zI - X) = \det(zI - gXg^{-1})$, the $\chi_i(X)$ are invariant. Also, since $(gXg^{-1})^{i-1} = gX^{i-1}g^{-1}$, we have

$$(Zg^{-1})(gXg^{-1})^{i-1}(gY) = Zg^{-1}gX^{i-1}g^{-1}gY = ZX^{i-1}Y.$$

Corollary 19.5. $\theta(X, Y, Z)$ *is an element of* R^G.

Corollary 19.6. $k[S_{1,1}^n]^G = (R^G)_\theta$.

Proof: If $f \in k[S_{1,1}^n]^G$, then $f = f_1/\theta^m$ with $m \geq 0$ and $f_1 \in k[X, Y, Z] = R$. Since $\theta^m f = f_1$ and $\theta^m f$ is invariant, $f_1 \in R^G$.

Now, what we have shown so far is that $k[\psi_1, \ldots, \psi_{2n}]$ is contained in R^G and hence, that $\psi^*(R_0) \simeq k[\mathrm{Char}(n, k)] \simeq k[\psi_1, \ldots, \psi_{2n}]_\theta$ is contained in $(R^G)_\theta = k[S_{1,1}^n]^G$. In other words, to establish the third property of a geometric quotient, namely that $k[S_{1,1}^n]^G \text{``=''} k[\mathrm{Char}(n, k)]$, it will be enough to prove the following.

Theorem 19.7. *Let* $S = R^G = k[X, Y, Z]^G$ *and* $S_0 = k[\psi_1, \ldots, \psi_{2n}]$ $= k[ZY, \ldots, ZX^{n-1}Y, \chi_1(X), \ldots, \chi_n(X)]$. *Then* $S_0 = S$.

We shall give several proofs of the theorem. The first involves a "canonical form". The second involves invariants under the symmetric group (Appendix C). A third proof, which actually deals with $\mathrm{Hank}(n, k)$, involves a classical determinantal approach. A fourth proof which will not be developed here involves invoking the Hilbert-Mumford Theory ([M-3], [M-4]).

We first note the following:

Proposition 19.8. *Let* $\omega(X, Y, Z) = \det \mathbf{Z}(X, Y, Z)$. *Then* $(\mathbb{A}_k^{n^2+2n})_\omega$ *is invariant under* G.

Proof: If $x = (A, b, c) \in (\mathbb{A}_k^{n^2+2n})_\omega$, then $\omega(A, b, c) \neq 0$ and $\omega(g \cdot x) = \omega(A, b, c) \cdot (\det g^{-1})$ (by proposition 11.13) so that $\omega(g \cdot x) \neq 0$.

Since $(\mathbb{A}_k^{n^2+2n})_\omega$ is an open dense G-invariant subset of $\mathbb{A}_k^{n^2+2n}$, an f in $S = R^G$ is determined by its restriction to this set (e.g. by corollary 12.16).

Proposition 19.9. *If* $X_c(X) = X_c$ *is the "companion matrix" given* *by*

$$X_c = \begin{bmatrix} 0 & 1 & 0 & \cdots & 0 \\ 0 & 0 & 1 & \cdots & 0 \\ \vdots & \vdots & \vdots & & \vdots \\ \vdots & \vdots & \vdots & 0 & 1 \\ -\chi_n(X) & -\chi_{n-1}(X) & \cdots & \cdots & -\chi_1(X) \end{bmatrix} \tag{19.10}$$

then

$$\begin{aligned} \mathbf{Z}(X,Y,Z)X &= X_c\mathbf{Z}(X,Y,Z) \\ \mathbf{Z}(X,Y,Z)Y &= (ZX^{j-1}Y)_{j=1}^n \\ \begin{bmatrix} 1 & 0 & \cdots & 0 \end{bmatrix} \mathbf{Z}(X,Y,Z) &= Z \end{aligned} \tag{19.11}$$

(as polynomial identities).

Proof: Since $\mathbf{Z}(X,Y,Z) = (ZX^{j-1})_{j=1}^n$ (cf. 11.2), we clearly have $\mathbf{Z}(X,Y,Z)Y = (ZX^{j-1}Y)_{j=1}^n$ and $\begin{bmatrix} 1 & 0 & \cdots & 0 \end{bmatrix} \mathbf{Z}(X,Y,Z) = Z$. Now if M is an $n \times n$ matrix, let M_j denote the j-th column of M. Thus, $X = [X_1, \ldots, X_N]$ and $X^r = [(X^r)_1 \cdots (X^r)_n] = [X^{r-1}X_1 \cdots X^{r-1}X_n]$. It follows that

$$\mathbf{Z}(X,Y,Z)X = \begin{bmatrix} ZX_1 & \cdots & ZX_n \\ ZXX_1 & \cdots & ZXX_n \\ \vdots & & \vdots \\ ZX^{n-1}X_1 & \cdots & ZX^{n-1}X_n \end{bmatrix}$$

$$= \begin{bmatrix} ZX_1 & \cdots & ZX_n \\ Z(X^2)_1 & \cdots & Z(X^2)_n \\ \vdots & & \vdots \\ z(X^n)_1 & \cdots & Z(X^n)_n \end{bmatrix} \tag{19.12}$$

But, by the Cayley-Hamilton theorem, $X^n = -\chi_1(X)X^{n-1} - \cdots - \chi_n(X)I$ and so, $(X^n)_j = -\chi_1(X)(X^{n-1})_j - \cdots - \chi_n(X)I_j$. The assertion that $\mathbf{Z}(X,Y,Z)X = X_c\mathbf{Z}(X,Y,Z)$ follows immediately.

Corollary 19.13. *If $(A,b,c) \in (\mathbb{A}_k^{n^2+2n})_\omega$, then*

$$\mathbf{Z}(A,b,c)A\mathbf{Z}(A,b,c)^{-1} = A_c,$$
$$c\mathbf{Z}(A,b,c)^{-1} = [1 \quad 0 \quad \cdots \quad 0],$$

and

$$\mathbf{Z}(A,b,c)b = (cA^{j-1}b)_{j=1}^{n}.$$

Corollary 19.14. *Let* $f(X,Y,Z)$ *be an element of* $R = k[X,Y,Z]$ *and set* $f_\omega(X,Y,Z) = f(X_c,(ZX^{j-1}Y)_{j=1}^{n},[1 \quad 0 \quad \cdots \quad 0])$. *If* f *is an element of* $S = R^G$, *then* $f = f_\omega$.

Proof: It is enough to show that $f = f_\omega$ on $(\mathbb{A}_k^{n^2+2n})_\omega$. But $(f - f_\omega)(A,b,c) = f(A,b,c) - f_\omega(A,b,c) = f(A,b,c) - f(gAg^{-1},gb,cg^{-1})$ for $(A,b,c) \in (\mathbb{A}_k^{n^2+2n})_\omega$ where $g = \mathbf{Z}(A,b,c)$ in view of the previous corollary. Since f is invariant, $f(A,b,c) = f(gAg^{-1},gb,cg^{-1})$ and the corollary follows.

Corollary 19.15. *If* $f \in S$, *then* $f = f_\omega \in S_0$.

Proof: Simply note that $f_\omega \in S_0$ and apply corollary 19.14.

In other words, we have shown that $S_0 = S$ and thus, have established theorem 19.7 so that the third property of a geometric quotient is satisfied. We note that a completely analogous proof can be obtained using $\gamma(X,Y,Z) = \det \mathbf{Y}(X,Y,Z)$ where $\mathbf{Y}$ is the controllability map. In particular, we have the following:

Proposition 19.16. $(\mathbb{A}_k^{n^2+2n})_\gamma$ *is G-invariant.*

Proposition 19.17.   *If* $X^c(X) = X^c$ *is the "companion matrix" given by*

$$X^c(X) = \begin{bmatrix} 0 & 0 & \cdots & 0 & -\chi_n(X) \\ 1 & 0 & \cdots & \vdots & -\chi_{n-1}(X) \\ 0 & 1 & \cdots & \vdots & \vdots \\ \vdots & \vdots & & 0 & \vdots \\ 0 & 0 & \cdots & 1 & -\chi_1(X) \end{bmatrix} \tag{19.18}$$

then

$$\mathbf{Y}(X,Y,Z)X^c = X\mathbf{Y}(X,Y,Z)$$
$$[ZX^{j-1}Y]_{j=1}^n = Z\mathbf{Y}(X,Y,Z)$$

$$\mathbf{Y}(X,Y,Z)\begin{bmatrix} 1 \\ 0 \\ \vdots \\ 0 \end{bmatrix} = Y \tag{19.19}$$

(as polynomial identities).

Corollary 19.20. *If* $(A,b,c) \in (\mathbb{A}_k^{n^2+2n})_\gamma$, *then*

$$\mathbf{Y}(A,b,c)^{-1}A\mathbf{Y}(A,b,c) = A^c,$$

$$\mathbf{Y}(A,b,c)^{-1}Y = \begin{bmatrix} 1 \\ 0 \\ \vdots \\ 0 \end{bmatrix},$$

and

$$c\mathbf{Y}(A,b,c) = [cA^{j-1}b]_{j=1}^n.$$

Corollary 19.21. *Let* $f(X,Y,Z)$ *be an element of* $R = k[X,Y,Z]$
and set $f_\gamma(X,Y,Z) = f(X^c, \begin{bmatrix} 1 \\ 0 \\ \vdots \\ 0 \end{bmatrix}, [ZX^{j-1}Y]_{j=1}^n)$. *If* f *is an element of*
$S = R^G$, *then* $f = f_\gamma$.

Corollary 19.22. *If* $f \in S$, *then* $f = f_\gamma \in S_0$.

We can interpret these results in terms of maps from $\mathbb{A}_k^{n^2+2n}$ into
$\mathbb{A}_k^{2n}$ and $\mathbb{A}_k^{2n}$ into $\mathbb{A}_k^{n^2+2n}$. More precisely, let $\alpha_1 : \mathbb{A}_k^{n^2+2n} \to \mathbb{A}_k^{2n}$ and
$\alpha_2 : \mathbb{A}_k^{2n} \to \mathbb{A}_k^{n^2+2n}$ be given by

$$\alpha_1(A,b,c) = (\mathbf{Z}(A,b,c)b, -\chi_1(A),\ldots,-\chi_n(A))$$
$$\alpha_2(h_1,\ldots,h_n,\chi_1,\ldots,\chi_n) = (A_c(\chi),h,[1 \quad 0 \quad \cdots \quad 0]) \tag{19.23}$$

where

$$A_c(\chi) = \begin{bmatrix} 0 & 1 & 0 & \cdots & 0 \\ 0 & 0 & 1 & \cdots & 0 \\ \vdots & & & & \vdots \\ \vdots & & & & 1 \\ -\chi_n & \cdots & \cdots & \cdots & -\chi_1 \end{bmatrix} \quad , \quad h = \begin{bmatrix} h_1 \\ h_2 \\ \vdots \\ h_n \end{bmatrix}$$

(note that $\alpha_1 = \mathfrak{R}_\chi$). Then it is easy to see that α_1, α_2 are morphisms and that $f \in S = R^G = k[X,Y,Z]^G$ if and only if $(\alpha_1 \circ \alpha_2)^*(f) = f$ (i.e. $f \circ \alpha_2 \circ \alpha_1 = f$). But $(\alpha_1 \circ \alpha_2)^*(f)$ is an element of S_0 and so, $S = S_0$.

We now give a second proof of the third property of a geometric quotient using the results of Appendix C and assuming that k has characteristic zero. We let $\Delta(A)$ be the discriminant of A so that $\Delta(X)$ is an element of $k[X,Y,Z]$. Then $\Delta(X)$ is an invariant under G (proposition C.19) so that $\Delta(X) \in S = R^G$. In fact, since $\Delta(X) = \Delta(\chi_1(X), \ldots, \chi_n(X))$, $\Delta(X) \in S_0$. We have:

Proposition 19.24. $(\mathbb{A}_k^{n^2+2n})_\Delta$ *is open and dense in* $\mathbb{A}_k^{n^2+2n}$ *and, consequently, a regular function f in S is determined by its restriction to* $(\mathbb{A}_k^{n^2+2n})_\Delta$.

Now let $D(n,k) \subset M(n,n;k)$ be the set of $n \times n$ diagonal matrices and let $D = \{(D(n,k),b,c)\}$ viewed as a subset of $\mathbb{A}_k^{n^2+2n}$. Then D is the affine variety in $\mathbb{A}_k^{n^2+2n}$ defined by the equations $X_{ij} = 0$, $i \neq j$. If $x = (A,b,c) \in (\mathbb{A}_k^{n^2+2n})_\Delta$, then there is a $D_x \, (= D_A)$ in $D(n,k)$ and a g in G with $gAg^{-1} = D_x$ such that $(D_x, gb, cg^{-1}) \in D \cap o_G(A,b,c)$. If f is a G-invariant regular function, then $f(A,b,c) = f(D_x, gb, cg^{-1})$ and so, f is determined by its restriciton to $(\mathbb{A}_k^{n^2+2n})_\Delta \cap D$. Since this set is open and dense in D, $f \in S$ is determined by its restriction f_D to D. In other words, S_0 will coincide with S i.e. Theorem 19.7 will hold, if we can prove the following:

Lemma 19.25. *If $f \in S$, then there is an $f_0 \in S_0$ such that both f and f_0 have the same restriction to D i.e. $f_D = f_{0D}$.*

We can immediately develop a proof of this lemma in the following way: let $\phi : \mathbb{A}_k^{n^2+2n} \to \mathbb{A}_k^{2n}$ be given by $\phi(A,b,c) = (c\mathbf{Y}(A,b,c),$ $-\chi_1(A),\ldots,-\chi_n(A))$ and, if $f \in R$, let $\tilde{f} : \mathbb{A}_k^{2n} \to \mathbb{A}_k^1$ be given by $\tilde{f}(u_1,\ldots,u_n,x_1,\ldots,x_n) = f(A_x,\epsilon_1,[u_1,\ldots,u_n])$ where $A_x = A_c(x)$

is an appropriate "companion matrix" and $\epsilon_1 = \begin{bmatrix} 1 \\ 0 \\ \vdots \\ 0 \end{bmatrix}$; if $f \in S =$

R^G, then $f(A,b,c) = f(A_c,\epsilon_1,c\mathbf{Y}(A,b,c)) = (\tilde{f} \circ \phi)(A,b,c)$ and so, $f(D_A,gb,cg^{-1}) = (\tilde{f} \circ \phi)(D_A,gb,cg^{-1})$ when A has distinct eigenvalues; but $(\tilde{f} \circ \phi)_D$ is an element of S_{oD} and lemma 19.25 follows. Of course, this approach is simply a variant on what we did before.

Now let us look at a different approach to the lemma. Let $N = S_n \times D_0(n,k)$ be the subgroup of G which stabilizes D (see appendix C). We note that the action of N on D is given by

$$(\sigma,\boldsymbol{\alpha}) \cdot (d,b,c) = (d_\sigma, \boldsymbol{\alpha} b_\sigma, c_\sigma \boldsymbol{\alpha}^{-1}) \qquad (19.26)$$

where $\sigma \in S_n$ is a permutation matrix, $\boldsymbol{\alpha} = \text{diag}[\alpha_1,\ldots,\alpha_n]$ with $\alpha_i \neq 0$ all i, $d = \text{diag}[d_1,\ldots,d_n]$, $d_\sigma = \text{diag}[d_{\sigma(1)},\ldots,d_{\sigma(n)}]$, $b = $ column $(b_1,\ldots,b_n)$, $b_\sigma = $ column $(b_{\sigma(1)},\ldots,b_{\sigma(n)})$, $c = $ row$(c_1,\ldots,$ $c_n)$ and $c_\sigma = $ row$(c_{\sigma(1)},\ldots,c_{\sigma(n)})$. A fortiori N acts on $k[D] = R_D$ which can be identified with the subring $k[X_{11},\ldots,X_{nn},Y,Z]$ of R. Moreover, if $f \in S = R^G$, then

$$f_D \in R_D^N = k[X_{11},X_{22},\ldots,X_{nn},Y,Z]^N.$$

If p is an element of R_D^N and $p = f_D$ is the restriction of an invariant f in S, then $p - f$ viewed as an element of R vanishes on D. In other words, $p - f \in I(D) = (X_{ij}), i \neq j$, and therefore,

$$p^g - p \in (X_{ij},X_{ij}^g), \qquad i \neq j \qquad (19.27)$$

for $g \in G$. However, since $N < G$, there may be functions in R_D^N which are not the restrictions of functions in S.

Example 19.28.   Let $n = 2$ and let $p(X_{11},X_{12},X_{21},X_{22},Y_1,Y_2,$ $Z_1,Z_2) = Z_1Y_1Z_2Y_2$. Then p is an element of R_D^N but p is not the restriction of an element of S. Let

$$g = \begin{bmatrix} 1 & 1 \\ 0 & 1 \end{bmatrix} , \quad g^{-1} = \begin{bmatrix} 1 & -1 \\ 0 & 1 \end{bmatrix}$$

so that

$$Y_1' = (gY)_1 = Y_1 + Y_2 \qquad Z_1' = (Zg^{-1})^1 = Z_1$$
$$Y_2' = (gY)_2 = Y_2 \qquad Z_2' = (Zg^{-1})^2 = Z_2 - Z_1$$

and $Z_1'Y_1'Z_2'Y_2' = Z_1(Y_1 + Y_2)(Z_2 - Z_1)Y_2 = Z_1Y_1Z_2Y_2 - Z_1Y_1Z_1Y_2 + Z_1Y_2Z_2Y_2 - Z_1Y_2Z_1Y_2 = p^g(X_{11}, X_{12}, X_{21}, X_{22}, Y_1, Y_2, Z_1, Z_2)$. If p were the restriction of an element of S, then

$$p^g - p = -Z_1Y_1Z_1Y_2 + Z_1Y_2Z_2Y_2 - Z_1Y_2Z_1Y_2$$

would be an element of the ideal $(X_{12}, X_{21}, X_{12}+X_{22}-X_{11}-X_{21}, X_{21})$ $= (X_{12}, X_{21}, X_{22} - X_{11})$ which is clearly false. Note, for example, that if $q(X, Y, Z) = Z_1Y_1 + Z_2Y_2$, then

$$q^g(X, Y, Z) = Z_1(Y_1 + Y_2) + (Z_2 - Z_1)Y_2 = Z_1Y_1 + Z_2Y_2 = q(X, Y, Z).$$

Example 19.29.   Let us now generalize the previous example. Suppose that $n > 2$ and set $W_i = Z_iY_i$ for $i = 1, \ldots, n$. Let $\boldsymbol{\psi}_1(W_1, \ldots, W_n)$, $\boldsymbol{\psi}_2(W_1, \ldots, W_n), \ldots, \boldsymbol{\psi}_n(W_1, \ldots, W_n)$ be the elementary symmetric polynomials in the W_i and observe that the $\boldsymbol{\psi}_j$ are elements of R_D^N (see appendix C). If $j \geq 2$, then $\boldsymbol{\psi}_j(W_1, \ldots, W_n)$ is not the restriction of an element of S. To see this, let $g \in G$ be given by

$$g = \begin{bmatrix} 1 & & 1 & & O_{2,n-2} \\ 0 & & 1 & & \\ & & & 1 & \\ O_{n-2,2} & & & & 1 \end{bmatrix}$$

where $O_{s,t}$ is an $s \times t$ matrix with 0 entries. Then $\boldsymbol{\psi}_j$ contains a *unique* term involving W_1W_2 and the remainder is invariant under g. Thus, $\boldsymbol{\psi}_j^g - \boldsymbol{\psi}_j$ is not an element of (X_{ij}, X_{ij}^g) and so $\boldsymbol{\psi}_j$ is not the restriction of an invariant to D. For example, if $n = 4$, then $\boldsymbol{\psi}_2(W) = W_1W_2 + W_1W_3 + W_1W_4 + W_2W_3 + W_2W_4 + W_3W_4 = W_1W_2 + (W_1 + W_2)(W_3 + W_4) + W_3W_4$, $\boldsymbol{\psi}_3(W) = W_1W_2W_3 + W_1W_2W_4 + W_1W_3W_4 + W_2W_3W_4 = (W_1W_2)(W_3 + W_4) + (W_1 + W_2)W_3W_4$ and $\boldsymbol{\psi}_4(W) = (W_1W_2)(W_3W_4)$ and $W_1 + W_2$ is invariant under g as are $W_3 + W_4$ and W_3W_4.

Bearing these examples in mind, we let $W_i = Z_iY_i$, $i = 1, 2, \ldots,$ n. If $\boldsymbol{\alpha} = \text{diag}[\alpha_1, \ldots, \alpha_n] \in D_0(n, k)$, then $W_i^{\boldsymbol{\alpha}} = Z_i\alpha_i^{-1}\alpha_iY_i =$

$Z_i Y_i = W_i$ so that the W_i are invariant under $D_0(n,k)$. Moreover, $X_{ii}^{\boldsymbol{\alpha}} = \alpha_i X_{ii} \alpha_i^{-1} = X_{ii}$ so that the X_{ii} are also invariant under $D_0(n,k)$. Set $X = (X_{11}, \ldots, X_{nn})$ and $W = (W_{11}, \ldots, W_{nn})$ so that S_n acts on $k[X, W] \subset R_D$ via $\sigma \cdot (X, W) = (X_\sigma, W_\sigma)$ where $X_\sigma = (X_{\sigma(1)\sigma(1)}, \ldots, X_{\sigma(n)\sigma(n)})$ and $W_\sigma = (W_{\sigma(1)\sigma(1)}, \ldots, W_{\sigma(n)\sigma(n)})$. Then $k[X, W]^{S_n} = k[X_{11}, \ldots, X_{nn}, W_{11}, \ldots, W_{nn}]^{S_n} \subset R_D^N$. In fact, we have:

Proposition 19.30. $R_D^N = k[X, W]^{S_n}$.

Proof: If $p \in R_D^N$, then $p = p_0 + p_1 + \cdots + p_m$ where p_i is homogeneous of degree i (or 0). Since the action of N preserves degrees, each $p_i \in R_D^N$ and so, we may suppose p homogeneous. Let $p = \Sigma a_{rst} X^r Z^s Y^t = \Sigma a_{rst} (X_{11})^{r_1} \cdots (X_{nn})^{r_n} Z_1^{s_1} \cdots Z_n^{s_n} Y_1^{t_1} \cdots Y_n^{t_n}$ with $a_{rst} \neq 0$. If $\boldsymbol{\alpha} = \mathrm{diag}[\alpha_1, \ldots, \alpha_n] \in D_0(n,k)$, then $p^{\boldsymbol{\alpha}} = \Sigma a_{rst} \alpha_1^{t_1 - s_1} \cdot \alpha_n^{t_n - s_n} (X_{11})^{r_1} \cdots (X_{nn})^{r_n} Z_1^{s_1} \cdots Z_n^{s_n} Y_1^{t_1} \cdots Y_n^{t_n}$. It follows that $p^{\boldsymbol{\alpha}} = p$ implies $s_1 = t_1, \ldots, s_n = t_n$ (for example, take $\alpha_2 = \cdots = \alpha_n = 1$ and equate coefficients in $p^{\boldsymbol{\alpha}} - p$ to obtain $(\alpha_1^{t_1 - s_1}) a_{rst} = 0$). In other words, $p \in k[X, W]^{S_n}$.

Now let $U_1 = (X_{11}, W_1), U_2 = (X_{22}, W_2), \ldots, U_n = (X_{nn}, W_n)$ so that $k[X, W] = k[U]$. Then S_n acts on $k[U]$ via

$$\sigma \cdot U_i = U_{\sigma(i)} = (X_{\sigma(i)\sigma(i)}, W_{\sigma(i)\sigma(i)}) \qquad (19.31)$$

and $k[X, W]^{S_n} = k[U]^{S_n}$. Writing $U_j = (U_{j^1}, U_{j^2})$ and letting $\mathbf{O}(p)$ be the sum of the elements in the orbit of p under S_n where p is a monomial in the U_j, we have:

Proposition 19.32. *Suppose that k has characteristic zero. Then $k[U]^{S_n}$ is generated by the multilinear terms $\mathbf{O}(U_{1i_1} \cdots U_{si_s}), s \leq n$.*

Proof: See Appendix C, corollary C.27.

Example 19.33. Let $n = 2$ so that $U_1 = (X_{11}, W_1)$ and $U_2 = (X_{22}, W_2)$. Then $k[X, W]^{S_n}$ is generated by the terms $\mathbf{O}(X_{11}) = X_{11} + X_{22}$, $\mathbf{O}(W_1) = W_1 + W_2$, $\mathbf{O}(X_{11} X_{22}) = 2 X_{11} X_{22}$, $\mathbf{O}(W_1 W_2) = W_1 W_2$, and $\mathbf{O}(X_{11} W_1) = X_{11} W_1 + X_{22} W_2$. We observe that $\mathbf{O}(X_{11}) = \chi_1(X)_D$, $\mathbf{O}(W_1) = (ZY)_D$, $\mathbf{O}(X_{11} X_{22}) = (2\chi_2(X))_D$, $\mathbf{O}(X_{11} W_1) = (ZXY)_D$ and that $W_1 W_2 \neq f_D$ for any f in S. Thus, if f is an invariant, then $f_D \in k[X_{11} + X_{22}, W_1 + W_2, X_{11}X_{22}, X_{11}W_1 + X_{22}W_2]$ and hence, there is a f_0 in S_0 with $f_D = f_{0D}$.

Now let $S_D = \{f_D : f \in S\}$ and let $S_{0D} = \{f_{0D} : f \in S_0\}$. Then $S_{0D} \subset S_D \subset k[X, W]^{S_n} = R_D^N$. We observe that S_{0D} is generated by the $2n$ terms $\mathbf{O}(X_{11})$, $\mathbf{O}(X_{11}^2)$, $\ldots$, $\mathbf{O}(X_{11}^n)$, $\mathbf{O}(W_1)$, $\mathbf{O}(X_{11}W_1), \ldots, \mathbf{O}(X_{11}^{n-1}W_1)$. Let $p_{ij} = \mathbf{O}(X_{11}^i W_1^j)$, $i < j$ be the remaining generators of $k[X, W]^{S_n}$ so that $R_D^N = S_{0D}[p_{ij}]$. However, application of equation 19.27 and the calculations of example 19.29, show that none of the p_{ij} are in S_D and in addition, that if $p \in S_D$, then $p \in S_{0D}$ which establishes lemma 19.25. In other words, the second proof of the third property of a geometric quotient is complete.

We now turn our attention to the third proof of the ring of invariants property of a geometric quotient. If $x = (A, b, c)$ is an element of $\mathbb{A}_k^{n^2+2n}$, we let $\mathbf{Y}_1 : \mathbb{A}_k^{n^2+2n} \to M(n, n+1)$ and $\mathbf{Z}^1 : \mathbb{A}_k^{n^2+2n} \to M(n+1, n)$ be the morphisms given by

$$\mathbf{Y}_1(x) = \mathbf{Y}_1(A, b, c) = [b \quad Ab \quad \cdots \quad A^n b] = [\mathbf{Y}(A, b, c) \quad A^n b]$$

$$\mathbf{Z}^1(x) = \mathbf{Z}^1(A, b, c) = \begin{bmatrix} c \\ cA \\ \vdots \\ cA^n \end{bmatrix} = \begin{bmatrix} \mathbf{Z}(A, b, c) \\ cA^n \end{bmatrix} \tag{19.34}$$

where $\mathbf{Y}, \mathbf{Z}$ are the controllability and observability matrices. Letting $M_{*r}(s, t)$ be the set of $s \times t$ matrices of rank r, we have:

Proposition 19.35. $\mathbf{Y}_1(x) \in M_{*n}(n, n+1)$ *if an only if x is controllable; and, $\mathbf{Z}^1(x) \in M_{*n}(n+1, n)$ if and only if x is observable.*

Now let $\varphi : \mathbb{A}_k^{n^2+2n} \to M(n+1, n) \times M(n, n+1)$ be the morphism given by

$$\varphi(x) = (\mathbf{Z}^1(x), \mathbf{Y}_1(x)) \tag{19.36}$$

Then $\varphi(S_{1,1}^n) \subset M_{*n}(n+1, n) \times M_{*n}(n, n+1)$. If $M \in M(s, t)$ and $i_1, \ldots, i_r, j_1, \ldots, j_r$ are indices, then $\det \left[M \mid \begin{smallmatrix} i_1 \cdots i_r \\ j_1 \cdots j_r \end{smallmatrix} \right]$ is the $r \times r$ minor of M with rows $i_1, \ldots, i_r$ and columns $j_1, \ldots, j_r$. We let $\alpha(M) = \det \left[M \mid \begin{smallmatrix} 1 \cdots n \\ 1 \cdots n \end{smallmatrix} \right]$ for M in $M(n+1, n)$ and $\beta(\widetilde{M}) = \det \left[\widetilde{M} \mid \begin{smallmatrix} 1 \cdots n \\ 1 \cdots n \end{smallmatrix} \right]$ for $\widetilde{M}$ in $M(n, n+1)$.

Proposition 19.37. $\mathbf{Z}^1(S_{1,1}^n) = (M(n+1,n))_\alpha$ and $\mathbf{Y}_1(S_{1,1}^n) = (M(n,n+1))_\beta$.

Proof: Enough to treat $\mathbf{Y}_1(S_{1,1}^n)$. Clearly, $\mathbf{Y}_1(S_{1,1}^n) \subset (M(n,n+1))_\beta$. If $[v_1 \cdots v_n \ w] \in M(n,n+1)_\beta$, then $\det[v_1 \ldots, v_n] \neq 0$ and $w = \sum_{j=1}^n a_j v_j$, $a_j \in k$. Let $g \in G$ be such that $g \cdot [v_1 \ldots v_n] = I$. Set

$$A^* = \begin{bmatrix} 0 & \cdots & 0 & a_1 \\ 1 & \cdots & 0 & \vdots \\ 0 & & \vdots & \vdots \\ \vdots & & \vdots & \vdots \\ 0 & \cdots & 1 & a_n \end{bmatrix} \quad , \quad b^* = \epsilon_1 \ , \quad c^* = \epsilon^n = \begin{pmatrix} 0 & \cdots & 0 & 1 \end{pmatrix}$$

so that $A^{*j}b^* = \epsilon_{j+1}$ for $j = 0,\ldots,n-1$ and $A^{*n}b^* = \sum_{j=1}^n a_j\epsilon_j$. Let $b = g^{-1}b^*$ and $A = g^{-1}A^*g$ and $c = c^*g$. Since $g \cdot [v_1 \cdots v_n] = I = [\epsilon_1 \cdots \epsilon_n]$, we can see that $b = g^{-1}\epsilon_1 = v_1$, $Ab = g^{-1}A^*gg^{-1}\epsilon_1 = g^{-1}\epsilon_2 = v_2,\ldots$, $A^j b = g^{-1}A^{*j}b^* = g^{-1}\epsilon_{j+1} = v_{j+1}$ for $j = 0,1,\ldots$, $n-1$ and $A^n b = g^{-1}A^{*n}b^* = \Sigma a_j g^{-1}\epsilon_j = \sum_{j=1}^n a_j v_j = w$. In other words, $\mathbf{Y}_1(A,b,c) = [v_1,\ldots,v_n \ w]$. Since

$$\det \mathbf{Z}(A,b,c) = \det \mathbf{Z}(A^*,b^*,c^*) \cdot \det g \neq 0,$$

(A,b,c) is in $S_{1,1}^n$.

Corollary 19.38. $\varphi(S_{1,1}^n) \subset M(n+1,n)_\alpha \times M(n,n+1)_\beta$.

Let $\psi : M(n+1,n) \times M(n,n+1) \to M(n+1,n+1)$ be the morphism given by

$$\psi(Z,Y) = ZY \tag{19.39}$$

If $\theta(M) = \det \begin{bmatrix} M | \begin{smallmatrix} 1 \cdots n \\ 1 \cdots n \end{smallmatrix} \end{bmatrix}$ for M in $M(n+1,n+1)$, then $\psi(M(n+1,n)_\alpha \times M(n,n+1)_\beta) \subset M_{*n}(n+1,n+1)_\theta$. We let $\mathcal{L}_{n+1}$ be the (linear*) variety in $M(n+1,n+1)$ defined by the equations

$$\begin{aligned} M_{ij} - M_{ji} &= 0 & i,j &= 1,\ldots,n+1 \\ M_{in+1} - M_{i+1n} &= 0 & i &= 1,\ldots,n \end{aligned} \tag{19.40}$$

---

* See definition 20.5.

and let $\mathcal{H} = \mathcal{L}_{n+1} \cap M_{*n}(n+1, n+1)_\theta$. $\mathcal{H}$ is a quasi-affine variety since it is an open subset of $\mathcal{L}_{n+1}$.

Proposition 19.41. $(\psi \circ \varphi)(S_{1,1}^n) = \mathcal{H}$.

Proof: If $x \in S_{1,1}^n$, then $(\psi \circ \varphi)(x) = \mathbf{Z}^1(x)\mathbf{Y}_1(x)$ clearly satisfies (19.40). If H is an element of $\mathcal{H}$, then there are $h_1, \ldots, h_{2n+1}$ and $a_1, \ldots, a_n$ in k such that

$$
H = \begin{bmatrix}
h_1 & \cdots & h_n & h_{n+1} \\
\vdots & & \vdots & \vdots \\
h_n & \cdots & h_{2n-1} & h_{2n} \\
h_{n+1} & \cdots & h_{2n} & h_{2n+1}
\end{bmatrix}
$$

and

$$
h_{n+j} = \sum_{\ell=1}^{n} a_\ell h_{\ell+j-1},
$$

for $j = 1, \ldots, n+1$. Since $\det \left[H \mid \begin{matrix} 1 \cdots n \\ 1 \cdots n \end{matrix} \right] \neq 0$, there is, by realization theory (theorem 10.18), an $x = (A, b, c)$ in $S_{1,1}^n$ such that $h_\ell = cA^{\ell-1}b$, $\ell = 1, \ldots, 2n+1(\cdots)$. In other words, $(\psi \circ \varphi)(x) = H$.

Corollary 19.42. $\varphi(S_{1,1}^n) = \psi^{-1}(\mathcal{H})$ *so that* $\varphi(S_{1,1}^n)$ *is a quasi-affine variety.*

Proof: Clearly, $\varphi(S_{1,1}^n) \subset \psi^{-1}(\mathcal{H})$. If $(Z, Y) \in \psi^{-1}(\mathcal{H})$, then $\psi(Z, Y) = ZY = H \in \mathcal{H}$ and, by the proposition, there is an x in $S_{1,1}^n$ with $\psi(\varphi(x)) = H$. Let

$$
Z = \begin{bmatrix} Z_n \\ z \end{bmatrix}, \quad Y = [Y_n \ \ y], \quad \mathbf{Z}^1(x) = \begin{bmatrix} \mathbf{Z}_n^1 \\ z^1 \end{bmatrix}, \quad \mathbf{Y}_1(x) = [\mathbf{Y}_{1n} \ y_1]
$$

Since $H = ZY = \mathbf{Z}^1(x)\mathbf{Y}_1(x)$, we have $Z_n Y_n = \mathbf{Z}_n^1 \mathbf{Y}_{1n}$, $zY_n = z^1 \mathbf{Y}_{1n}$, $Z_n y = \mathbf{Z}_n^1 y_1$, and $zy = z^1 y_1$. But $Z_n Y_n = \mathbf{Z}_n^1 \mathbf{Y}_{1n}$ is non-singular so that $g = Y_n \mathbf{Y}_{1n}^{-1} \in G = GL(n, k)$ and $Z = \mathbf{Z}^1(x)g^{-1}$, $Y = g\mathbf{Y}_1(x)$ (as $Z_n^{-1}\mathbf{Z}_n^1 = Y_n \mathbf{Y}_{1n}^{-1}$). If $\tilde{x} = g \cdot x$, then $\varphi(\tilde{x}) = (\mathbf{Z}^1(\tilde{x}), \mathbf{Y}_1(\tilde{x})) = (\mathbf{Z}^1(x)g^{-1}, g\mathbf{Y}_1(x)) = (Z, Y)$. In other words, we have $\psi^{-1}(\mathcal{H}) \subset \varphi(S_{1,1}^n)$.

Set $\tilde{S}_{1,1}^n = \varphi(S_{1,1}^n)$. φ is a morphism and as a map of $S_{1,1}^n$ into $\tilde{S}_{1,1}^n$ is obviously surjective.

Proposition 19.43. $\varphi : S_{1,1}^n \to \tilde{S}_{1,1}^n$ is injective.

Proof: Suppose that $\varphi(x) = \varphi(x_1)$ where $x = (A, b, c)$ and $x_1 = (A_1, b_1, c_1)$. Then $b = b_1$, $c = c_1$, and (say) $\mathbf{Y}_1(A, b, c) = \mathbf{Y}_1(A_1, b, c)$ so that $A^j b = A_1^j b$, $j = 0, \ldots, n$. Since $v_j = A^j b$, $j = 0, \ldots, n-1$ is a basis of k^n, to show that $A = A_1$, it will be enough to show that $A v_j = A_1 v_j$, $j = 0, \ldots, n-1$. But $A v_j = A^{j+1} b = A_1 \cdot A_1^j b = A_1 v_j$ for $j = 0, 1, \ldots, n-1$ since $\mathbf{Y}_1(A, b, c) = \mathbf{Y}_1(A_1, b, c)$.

Proposition 19.44. $\varphi : S_{1,1}^n \to \tilde{S}_{1,1}^n$ is an isomorphism.

Proof: Since φ is bijective, we need only show that φ^{-1} is a morphism. If $(\mathbf{Z}^1(x), \mathbf{Y}_1(x)) = \varphi(x) \in \tilde{S}_{1,1}^n$, then $\varphi^{-1}(\mathbf{Z}^1(x), \mathbf{Y}_1(x)) = x = (A, b, c)$ and $b = (\mathbf{Y}_1(x))_1$, $c = (\mathbf{Z}^1(x))^1$. Clearly, b and c are regular and so, we need only show that $A(\mathbf{Z}^1(x), \mathbf{Y}_1(x))$ is regular. Let (say) $\mathbf{Y}_1(x) = [Y_1 \; \cdots \; Y_n \; Y_{n+1}]$ with $Y_{n+1} = a_1 Y_1 + \cdots + a_n Y_n$ and let $g \in G$ be $([Y_1 \cdots Y_n])^{-1}$. Then, g, g^{-1} are regular functions of $Y_1, \ldots, Y_n$ on $\tilde{S}_{1,1}^n$ and $A = g^{-1} A^* g$ where

$$A^* = \begin{bmatrix} 0 & 0 & \cdots & 0 & a_1 \\ 1 & 0 & \cdots & 0 & a_2 \\ \vdots & \vdots & \vdots & \vdots & \vdots \\ 0 & 1 & \cdots & 1 & a_n \end{bmatrix}$$

Thus, it suffices to show that the a_i are regular. But

$$a_j = \det[\epsilon_1 \; \cdots \; \epsilon_{j-1} \; g Y_{n+1} \; \epsilon_{j+1} \; \cdots \; \epsilon_n]$$

is clearly regular and so, φ^{-1} is a morphism.

Now we observe that G acts on $M(n+1, n) \times M(n, n+1)$ via

$$g \cdot (Z, Y) = (Z g^{-1}, g Y) \tag{19.45}$$

and that $\tilde{S}_{1,1}^n$ is invariant under this action. However, in view of proposition 11.13, $\varphi(g \cdot x) = g \cdot \varphi(x)$ so that φ is a G-isomorphism. Thus,

Corollary 19.46. $k[S_{1,1}^n]^G = k[\tilde{S}_{1,1}^n]^G$.

Hence, in order to establish the ring of invariants property of a geometric quotient, we need only prove the following:

Theorem 19.47. $k[\tilde{S}_{1,1}^n]^G = k[\mathcal{H}]$

Since $k[\mathcal{H}] \simeq \psi^*(k[\mathcal{H}]) \subset k[\tilde{S}_{1,1}^n]$ and $\psi(g \cdot (Z,Y)) = \psi((Z,Y))$, it is clear that $k[\mathcal{H}] \subset k[\tilde{S}_{1,1}^n]^G$ and so, we must show that $k[\tilde{S}_{1,1}^n]^G \subset k[\mathcal{H}]$ (or, more precisely, $\psi^*(k[\mathcal{H}])$). Let $V \subset \tilde{S}_{1,1}^n$ be the set of elements of the form

$$v = \left(\begin{pmatrix} I \\ z \end{pmatrix}, (\mathbf{Y}, y) \right) \tag{19.48}$$

i.e. $\mathbf{Z}(x) = I$. We consider $G \times V$ and we observe that G acts on $G \times V$ via

$$g \cdot (g_1, v) = (gg_1, v) \tag{19.49}$$

We have:

Lemma 19.50. $k[G \times V]^G = k[V]$

Proof: Let $R = k[G]$, $S = k[V]$. Then $R^G = k[G]^G = k$. It follows from lemma B.14 (say), that $(R \otimes_{R^G} S)^G = S$ i.e. that $(k[G] \otimes_k k[V])^G = (k[G \times V])^G = k[V]$. This can also be shown directly. If $f = \Sigma r_i \otimes s_i \in k[G] \otimes_k k[V]$, then $f(g,v) = \Sigma r_i(g)s_i(v)$. If f is invariant, then $f(g,v) = f(g \cdot I, v) = f(I,v) = \Sigma r_i(I)s_i(v)$ so that $f = \Sigma r_i(I)s_i \in k[V]$.

Define a morphism $\gamma : G \times V \to \tilde{S}_{1,1}^n$ by setting

$$\gamma((g,v)) = g \cdot v \tag{19.51}$$

Then:

Lemma 19.52. γ is a G-invariant isomorphism and so, $k[\tilde{S}_{1,1}^n]^G = k[G \times V]^G = k[V]$.

Proof: Clearly $\gamma(g_1 \cdot (g,v)) = \gamma((g_1 g, v)) = (g_1 g) \cdot v = g_1 \cdot (gv) = g_1 \cdot \gamma((g,v))$ so that γ is G-invariant. Define $\gamma^{-1} : \tilde{S}_{1,1}^n \to G \times V$ as follows:

$$\gamma^{-1}\left(\begin{pmatrix} \mathbf{Z} \\ z \end{pmatrix}, (\mathbf{Y}, y)\right) = \left(\mathbf{Z}^{-1}, \begin{pmatrix} I \\ z\mathbf{Z}^{-1} \end{pmatrix}, (\mathbf{ZY}, \mathbf{Z}y)\right)$$

so that (obviously) $\gamma \cdot \gamma^{-1} =$ identity and $\gamma^{-1} \circ \gamma =$ identity.

Corollary 19.53. *If f is an element of $k[\tilde{S}^n_{1,1}]^G$, then $f \in k[z\mathbf{Z}^{-1}, \mathbf{ZY}, \mathbf{Z}y]$ (i.e. f is a function of $z\mathbf{Z}^{-1}$, $\mathbf{ZY}$ and $\mathbf{Z}y$).*

In view of this corollary, we need only prove the following:

Lemma 19.54. $k[z\mathbf{Z}^{-1}, \mathbf{ZY}, \mathbf{Z}y] \subset k[\mathcal{H}]$.

Proof: Since $k[\mathcal{H}] = k[\mathbf{ZY}, \mathbf{Z}y, z\mathbf{Y}, zy]_{\det(\mathbf{ZY})}$, it is enough to show that $z\mathbf{Z}^{-1}$ is in $k[\mathcal{H}]$. But $\mathbf{Z}^{-1} = \mathbf{Y}(\mathbf{ZY})^{-1} = \mathbf{Y}\mathrm{adj}(\mathbf{ZY})/\det(\mathbf{ZY})$ and so $z\mathbf{Z}^{-1} = (z\mathbf{Y})\mathrm{adj}(\mathbf{ZY})/\det(\mathbf{ZY})$ is in $k[\mathcal{H}]$.

We also note that, in view of proposition 19.41 and the results of section 10, it is quite easy to show that $\mathrm{Hank}(n, k)$ and $\mathcal{H}$ are isomorphic (Exercise 56.).

A fourth proof using the Hilbert-Mumford theory can also be given but we shall not do so here.

20. Affine Algebraic Geometry: Simple Points

We have shown that $\text{Char}(n,k)$ (or $\text{Hank}(n,k)$ or $\text{Rat}(n,k)$) is a quasi-affine variety of dimension $2n$ which is a geometric quotient of $S_{1,1}^n$ modulo $GL(n,k)$. We shall now show, that, in addition, $\text{Hank}(n,k)$ or $\text{Char}(n,k)$ or $\text{Rat}(n,k)$ is what is called *nonsingular* i.e. has no singular points. This will, of course, again require some algebra.

Definition 20.1.   A mapping $\alpha : \mathbb{A}_k^N \to \mathbb{A}_k^N$ is called an *affine transformation* if there is a g in $GL(n,k)$ and a ξ_0 in $\mathbb{A}_k^N$ such that

$$\alpha(\xi) = g\xi + \xi_0 \qquad (20.2)$$

for all ξ in $\mathbb{A}_k^N$. If $g = I$ (the identity), then α is called a *translation*.

Let $A(N,k)$ be the set of all affine transformations of $\mathbb{A}_k^N$ and write $\alpha = (g, \xi_0)$ for α in $A(N,k)$. If $\alpha_1 = (g_1, \xi_1)$ and $\alpha_2 = (g_2, \xi_2)$, then $(\alpha_1\alpha_2)(\xi) = (g_1 g_2)\xi + g_1\xi_2 + \xi_1$ and we may define a multiplication in $A(N,k)$ by setting

$$(g_1, \xi_1)(g_2, \xi_2) = (g_1 g_2, g_1\xi_2 + \xi_1) \qquad (20.3)$$

Then $A(N,k)$ is an algebraic group with identity $i = (I, 0)$ and $(g, \xi_0)^{-1} = (g^{-1}, -g^{-1}\xi_0)$ (see exercise 57.). Since $A(N,k)$ acts on $\mathbb{A}_k^N$ via (20.2), $A(N,k)$ also acts on $k[X] = k[X_1, \ldots, X_N]$ via

$$(g, \xi_0) \cdot X = (g, \xi_0) \begin{bmatrix} X_1 \\ X_2 \\ \vdots \\ X_N \end{bmatrix} = g \begin{bmatrix} X_1 \\ X_2 \\ \vdots \\ X_N \end{bmatrix} + \xi_0 \qquad (20.4)$$

and this action preserves degrees. If $Z_i = (gX + \xi_0)_i$, $i = 1, \ldots, N$, then the Z_i are transcendental over k and $k[Z_1, \ldots, Z_N] = k[X_1, \ldots, X_N]$. If $V \subset \mathbb{A}_k^N$ is an affine algebraic set with $I(V) = (f_1(X), \ldots, f_r(X))$, then $\alpha(V)$ is also an affine algebraic set with $I(\alpha(V)) = \alpha(I(V))$. Moreover, if V is an affine variety so that $I(V)$ is prime, then $\alpha(V)$ is also an affine variety and $k[V] \simeq k[\alpha(V)]$ so that $\dim_k V = \dim_k \alpha(V)$.

Definition 20.5.   Let $\ell_1, \ldots, \ell_m$ be independent linear forms (homogeneous polynomials of degree 1) in $k[X_1, \ldots, X_N]$ and let $a_1, \ldots, a_m$ be elements of k. Then the affine algebraic set $V(\ell_1 - a_1, \ldots, \ell_m - a_m)$ is called a *linear variety*.

Proposition 20.6.   *If $V = V(\ell_1 - a_1, \ldots, \ell_m - a_m)$ is a linear variety in $\mathbb{A}_k^N$, then V is irreducible, $\dim_k V = N - m$, and V is isomorphic to a subspace of $\mathbb{A}_k^N$.*

Proof: Since $\ell_1, \ldots, \ell_m$ are independent, there is an affine transformation $\alpha = (g, \xi_0)$ such that if $Z_i = (gX + \xi_0)_i$, then $Z_i = \ell_i(X) - a_i$, $i = 1, \ldots, m$. In other words, $\alpha(V) = V(Z_1, \ldots, Z_m)$. But $V(Z_1, \ldots, Z_m)$ is a subspace of $\mathbb{A}_k^N$ and is irreducible and has dimension $N - m$. Since V and $\alpha(V)$ are isomorphic, the proposition is established.

Definition 20.7.   Let V be an affine variety so that $\mathfrak{p} = I(V)$ is prime and let $\xi = (\xi_1, \ldots, \xi_N) \in V$. Then the (Zariski) *tangent space to V at ξ* is the linear variety defined by

$$\sum_{i=1}^{N} \frac{\partial f}{\partial X_i}(\xi)(X_i - \xi_i) = 0 \tag{20.8}$$

for all $f \in \mathfrak{p}$. We denote this variety by $T_{V,\xi}$.

We observe that, in view of proposition 20.6, $T_{V,\xi}$ may be viewed as a (vector) subspace of $\mathbb{A}_k^N$ with origin ξ. We also note that if $f_1, \ldots, f_r$ are a basis of $\mathfrak{p} = I(V)$ (i.e. $\mathfrak{p} = (f_1, \ldots, f_r)$), then

$$\dim_k T_{V,\xi} = N - \text{rank}[J(f_1, \ldots, f_r; X_1, \ldots, X_N)(\xi)] \tag{20.9}$$

where $J(f_1, \ldots, f_r; X_1, \ldots X_N)(\xi) = \left(\frac{\partial f_i}{\partial X_j}(\xi) \right)$ is the Jacobian matrix of $f_1, \ldots, f_r$ with respect to $X_1, \ldots, X_N$ at ξ. We then have:

Proposition 20.10.   *If $V = V(\ell_1 - a_1, \ldots, \ell_m - a_m)$ is a linear variety, then $\dim_k T_{V,\xi} = \dim_k V$ for all ξ in V.*

Proposition 20.11.   *If V is an affine variety and ν is an integer, then $\{\xi \in V : \dim_k T_{V,\xi} \geq \nu\}$ is Zariski closed in V.*

Proof: In view of (20.9), $\{\xi \in V : \dim_k T_{V,\xi} \geq \nu\}$ is the zero set of $\mathfrak{p}$ and the ideal of $N - \nu + 1 \times N - \nu + 1$ minors of $J(f_1, \ldots, f_r; X_1, \ldots, X_N)$.

We now make the following:

Definition 20.12.   Let V be an affine variety and let $\xi \in V$. We say that ξ is a *simple point of V* if $\dim_k T_{V,\xi} = \dim_k V$ and that V is *nonsingular* or *smooth* if all points of V are simple.

Corollary 20.13. *A linear variety is nonsingular.*

We shall show that the tangent space can be defined *intrinsically* and that the definition is "local". We require the results on derivations developed in Appendix D.

Definition 20.14.   Let R be a k-algebra and let M be an R-module. A *k-derivation of R into M* is a k-linear map $D : R \to M$ such that (i) $D(a) = 0$ for all $a \in k$ and (ii) $D(r_1 r_2) = r_1 D(r_2) + r_2 D(r_1)$ for all $r_1, r_2 \in R$.

We write $\mathrm{Der}_k(R, M)$ for the set of all k-derivations of R into M and note that $\mathrm{Der}_k(R, M)$ is an R-module in a natural way. If $\phi : R \to S$ is a homomorphism of k-algebras and N is an S-module, then N can be viewed as an R-module by setting

$$r \cdot \mathbf{n} = \phi(r)\mathbf{n} \tag{20.15}$$

for $r \in R$ and $\mathbf{n} \in N$ and there is a natural homomorphism $\tilde{\phi} : \mathrm{Der}_k(S, N) \to \mathrm{Der}_k(R, M)$ of R-modules given by

$$\tilde{\phi}(D) = D \circ \phi \tag{20.16}$$

for $D \in \mathrm{Der}_k(S, N)$.

Now let V be an affine variety and let ξ be an element of V. We set $R = k[V]$ and $K = K(R) = k(V)$. If $M_\xi = I(\xi)$ is the maximal ideal of ξ in $k[X_1, \ldots, X_N] = k[X]$ and $\mathfrak{m}_\xi = M_\xi/I(V)$ is the maximal ideal of ξ in R, then $k = R/\mathfrak{m}_\xi = k[X]/M_\xi$ so that k may be viewed as an R-module (via the homomorphism $R \to R/\mathfrak{m}_\xi$) or as a $k[X]$-module (via the homomorphism $k[X] \to k[X]/M_\xi$). We may thus consider the spaces of derivations $\text{Der}_k(R, k)$ and $\text{Der}_k(k[X], k)$ at ξ. In that case, we write $\text{Der}_k(R, k_\xi)$ and $\text{Der}_k(k[X], k_\xi)$ and speak of derivations *centered at* ξ.

Proposition 20.17. $T_{V,\xi} = \text{Der}_k(R, k_\xi) = \{D' \in \text{Der}_k(k[X], k_\xi) : D'(I(V)) = 0\}$. *Hence, if* $V \subset W$, *then* $T_{V,\xi} \subset T_{W,\xi}$.

Proof: Clearly a derivation D in $\text{Der}_k(R, k_\xi)$ is the same thing as a derivation D' in $\text{Der}_k(k[X], k_\xi)$ such that $D'(\mathfrak{p}) = 0$ where $\mathfrak{p} = I(V)$. If $D' \in \text{Der}_k(k[X], k_\xi)$, then D' is determined by its values $a_i = D'(X_i)$, $i = 1, \ldots, N$, and conversely. Since $D'f = \sum_{i=1}^N \frac{\partial f}{\partial X_i}(\xi)D'X_i$ for all f in $k[X]$, the derivations D' such that $D'(\mathfrak{p}) = 0$ correspond to values $a_1, \ldots, a_N$ such that $\sum_{i=1}^N \frac{\partial f}{\partial X_i}(\xi)a_i = 0$ for all f in $\mathfrak{p}$ and the proposition holds. (The last assertion follows from $I(W) \subset I(V)$.).

If $\psi : V \to W$ is a morphism and $\psi^* : k[W] \to k[V]$ is the comorphism of ψ, then there is an induced homomorphism $(d\psi)_\xi : T_{V,\xi} \to T_{W,\psi(\xi)}$ of the tangent spaces given by

$$(d\psi)_\xi(D) = D \circ \psi^* \tag{20.18}$$

where $D \in \text{Der}_k(k[V], k_\xi)$. We note that $(d\psi)_\xi$ is a linear map which we call the *tangent map at* ξ.

Lemma 20.19. *Let* $\psi : V \to W$ *be an isomorphism of affine varieties. Then* $(d\psi)_\xi$ *is an isomorphism of* $T_{V,\xi}$ *onto* $T_{W,\psi(\xi)}$.

Proof: Let $R = k[V]$, $S = k[W]$ and let $\psi^* : S \to R$ be the comorphism of ψ. We let $\overline{\psi}^* : S/\mathfrak{m}_{\psi(\xi)} \to R/\mathfrak{m}_\xi$ be the map given by

$$\overline{\psi}^*(\overline{f}) = \overline{\psi^*(f)} = \overline{(f \circ \psi)} \tag{20.20}$$

for $f \in S$. If $g \in \mathfrak{m}_\xi$, then there is an f in S with $\psi^*(f) = g$ and so, $\mathfrak{m}_\xi \subset \psi^*(\mathfrak{m}_{\psi(\xi)})$. Conversely, if $f \in \mathfrak{m}_{\psi(\xi)}$, then $\psi^*(f)(\xi) = f(\psi(\xi)) = 0$ and so, $\overline{\psi}^*$ is well-defined. Clearly $\overline{\psi}^*$ is an isomorphism and so, k_ξ " $=$ " $k_{\psi(\xi)}$. We write $\tilde{\psi}^*$ for $(d\psi)_\xi$ so that $(\tilde{\psi}^*)(D) = D \circ \psi^*$ for

$D \in \mathrm{Der}_k(R, k_\xi)$. If $\tilde{\psi}^*(D) = 0$, then $(D \circ \psi^*)(f) = 0$ for all f in S and therefore, $D(g) = 0$ for all g in R as ψ^* is an isomorphism. In other words, $\tilde{\psi}^*$ is injective. On the other hand, if $\tilde{D} \in \mathrm{Der}_k(S, k_{\psi(\xi)})$, then $D = \tilde{D} \circ \psi^{*-1} \in \mathrm{Der}_k(R, k_\xi)$ and $\tilde{\psi}^*(D) = \tilde{D}$ so that $\tilde{\psi}^*$ is surjective. Thus, the lemma is established.

Corollary 20.21. *Let $\psi : V \to W$ be an isomorphism of affine varieties. If V is nonsingular, then W is nonsingular.*

Corollary 20.22. *Let V and W be affine varieties and let ψ be an isomorphism of V onto an open subvariety of W. If $\xi \in V$, then $(d\psi)_\xi$ is an isomorphism of $T_{V,\xi}$ onto $T_{W,\psi(\xi)}$.*

Proof: Since the principal affine open sets form a base for the Zariski topology, we may, in view of the lemma, assume that $V = W_f$ for $f \in k[W]$ and that $\psi : W_f \to W$ is the natural injection. Then $k[W_f] = k[W][Y]/(1 - fY)$ and a derivation D in $\mathrm{Der}_k(k[W], k_\xi)$ extends to a derivation D' in $\mathrm{Der}_k(k[W_f], k_\xi)$ by setting $D'(\overline{Y}) = -(Df)(\xi)/f(\xi)^2$. The result follows easily.

Corollary 20.23. *If V is a principal affine open subset of a linear variety, then V is nonsingular.*

Corollary 20.24. $\mathrm{Char}(n, k)$, $\mathrm{Hank}(n, k)$ *and* $\mathrm{Rat}(n, k)$ *are all nonsingular.*

We shall soon use corollary 20.23 to show that $\mathrm{Hank}(n, k)$ is nonsingular in a different way than simply observing that it is a principal affine open subset of $\mathbb{A}_k^{2n}$. Before doing so, we shall show that a product of nonsingular varieties is nonsingular and that the singular points (i.e. non simple) lie in a closed set.

Proposition 20.25.   *Let V and W be affine varieties and let (ξ, η) be an element of $V \times W$. Then*

$$T_{V,\xi} \times T_{W,\eta} \simeq T_{V \times W,(\xi,\eta)} \tag{20.26}$$

and therefore, if ξ is simple on V and η is simple on W, then (ξ, η) is simple on $V \times W$.

Proof: Let α, β be the natural isomorphisms of V, W into $V \times W$ given by

$$\alpha(v) = (v, \eta) \quad , \quad \beta(w) = (\xi, w) \tag{20.27}$$

so that

$$\alpha^*(f \otimes g) = fg(\eta) \quad , \quad \beta^*(f \otimes g) = f(\xi)g \tag{20.28}$$

for the comorphisms. We note that α^*, β^* are surjective. We define a map $(d\alpha)_\xi \times (d\beta)_\eta : T_{V,\xi} \times T_{W,\eta} \to T_{V \times W, (\xi, \eta)}$ by setting

$$(d\alpha)_\xi \times (d\beta)_\eta(D, D') = D \circ \alpha^* + D' \circ \beta^* \tag{20.29}$$

for derivations D, D'. If $D \circ \alpha^* + D' \circ \beta^* = 0$ and f is an element of $k[V]$, then $f \otimes 1$ is an element of $k[V \times W]$, $\alpha^*(f \otimes 1) = f$ and $[D \circ \alpha^* + D' \circ \beta^*](f \otimes 1) = D(f) = 0$. In other words, $D = 0$. Similarly, $D' = 0$ and so, $(d\alpha)_\xi \times (d\beta)_\eta$ is injective. Now suppose $\tilde{D}$ is an element of $\mathrm{Der}_k(k[V \times W], k_{(\xi, \eta)})$. Set $D(f) = \tilde{D}(f \otimes 1)$ for $f \in k[V]$ and $D'(g) = \tilde{D}(1 \otimes g)$ for $g \in k[W]$. Then $\tilde{D}(f \otimes g) = \tilde{D}((f \otimes 1)(1 \otimes g)) = \tilde{D}(f \otimes 1)(1 \otimes g)(\xi, \eta) + (f \otimes 1)(\xi, \eta)\tilde{D}(1 \otimes g) = \tilde{D}(f \otimes 1)g(\eta) + f(\xi)\tilde{D}(1 \otimes g) = (D \circ \alpha^*)(f \otimes g) + (D' \circ \beta^*)(f \otimes g) = [D \circ \alpha^* + D' \circ \beta^*](f \otimes g)$ so that $(d\alpha)_\xi \times (d\beta)_\eta$ is surjective. As for the final assertion, simply note that $\dim_k(T_{V,\xi} \times T_{W,\eta}) = \dim_k T_{V,\xi} + \dim_k T_{W,\eta}$.

We now make the following

Definition 20.30.   Let $\mathrm{Sing}(V) = \{\xi \in V : \xi \text{ is singular}\}$. $\mathrm{Sing}(V)$ is called the *singular locus of* V.

We shall show that $\mathrm{Sing}(V)$ is a proper closed subset of V.

Proposition 20.31.   *Let V be an affine variety. Then there is a nonempty open subset U of V such that $\dim_k T_{V,\xi} = \dim_k V$ for ξ in U.*

Proof ([M-2]). Now $\dim_k V = \mathrm{tr.\ deg}\, K/k$ where $k = k(V)$ is the function field of V. Since K is separably generated over k (Appendix D), we have $\mathrm{tr.\ deg}\, K/k = \dim_k \mathrm{Der}_k(K, K) = \dim_K \mathrm{Der}_k(k[V], K)$. But $\mathrm{Der}_k(k[V], K) = \{D' \in \mathrm{Der}_k(k[X_1, \ldots, X_N], K) : D'(\mathfrak{p}) = 0\}$ where $\mathfrak{p} = I(V)$. If $\mathfrak{p} = (f_1, \ldots, f_r)$, then we have $\dim_K \mathrm{Der}_k(k[V], K)$ as the dimension over K of the subspace of $\mathbb{A}_K^N$ defined by the vanishing of the terms $\sum_{i=1}^N \frac{\partial f_j}{\partial X_i} X_i, j = 1, \ldots, r$. Thus, it will be enough to show that rank $J(f_1, \ldots, f_r; X_1, \ldots, X_N)$ in K is the same as rank $J(f_1, \ldots, f_r; X_1, \ldots, X_N)(\xi)$ for all ξ in an open subset U of V. If ρ is the rank of $J(f_1, \ldots, f_r; X_1, \ldots, X_N)$ in K, then there are $P \in GL(r, k)$ and $Q \in GL(N, K)$ such that

$$P\overline{J}(f_1, \ldots, f_r; X_1, \ldots X_N)Q = \begin{bmatrix} I_\rho & 0 \\ 0 & 0 \end{bmatrix}$$

where $\overline{J}(f_1, \ldots, f_r; X_1, \ldots, X_N) = J(f_1, \ldots, f_r; X_1, \ldots, X_N)$ modulo p. Let $P = P_0/p_0$, $Q = Q_0/q_0$ where P_0, Q_0 have entries in $k[V]$ and $p_0, q_0 \in k[V]$. Let $f = p_0 q_0 \det P_0 \det Q_0$ and $U = V_f$ (a principal affine open subset of V). If $\xi \in U$, then

$$P_0(\xi)J(f_1, \ldots, f_r; X_1, \ldots, X_N)(\xi)Q_0(\xi) = p_0(\xi)\begin{bmatrix} I_\rho & 0 \\ 0 & 0 \end{bmatrix}q_0(\xi)$$

so that $\operatorname{rank} J(f_1, \ldots, f_r; X_1, \ldots X_N)(\xi) = \rho$ and the proposition holds.

Corollary 20.32. $\dim_k T_{V,\xi} \geq \dim_k V$ *for all* $\xi \in V$.

Proof: Since $\{\xi : \dim_k T_{V,\xi} \geq \dim_k V\}$ is closed by proposition 20.11, its complement is open and hence, must be empty in view of the proposition and the irreducibility of V.

Corollary 20.33. $\operatorname{Sing}(V)$ *is closed (as* $\operatorname{Sing}(V) = \{\xi : \dim_k T_{V,\xi} \geq \dim_k V + 1\}$*).*

Now let us turn our attention to $\operatorname{Hank}(n, k)$. Let $\operatorname{Sym}(n, k)$ be the set of $n \times n$ symmetric matrices with entries in k. Then $\operatorname{Sym}(n, k)$ is a linear variety in $\mathbb{A}_k^{n^2}$ as it is defined by the equations $X_{ij} - X_{ji} = 0$ for $i \neq j$. Moreover, $\dim_k \operatorname{Sym}(n, k) = n(n+1)/2$. We identify $\operatorname{Sym}(n, k)$ with $\mathbb{A}_k^{n(n+1)/2}$ by using the coordinate functions X_{ij}, $i \geq j$, $i, j = 1, \ldots, n$. Let $H(n, k)$ be the subset of $\operatorname{Sym}(n, k)$ consisting of matrices of the form

$$H = \begin{bmatrix} h_1 & h_2 & h_3 & \cdots & h_n \\ h_2 & h_3 & h_4 & \cdots & h_{n+1} \\ \vdots & \vdots & \vdots & & \vdots \\ h_n & h_{n+1} & h_{n+2} & \cdots & h_{2n-1} \end{bmatrix} \tag{20.34}$$

where $h_1, \ldots, h_{2n-1}$ are elements of k. Since $\mathrm{Hank}(n,k)$ is isomorphic with the set $H(n,k)_{\det H} \times A_k^1$, it will be enough, in view of corollary 20.23, to show that $H(n,k)$ is a linear variety in $\mathrm{Sym}(n,k)$ in order to prove that $\mathrm{Hank}(n,k)$ is nonsingular.

Example 20.35.   Let $n = 3$. Then $H(3,k)$ is defined by the equation $X_{13} - X_{22} = 0$ in $\mathrm{Sym}(3,k)$ and so $H(3,k)$ is a linear variety of dimension 5.

Example 20.36.   Let $n = 4$. Then $H(4,k)$ is defined by the equations $X_{13} - X_{22} = 0$, $X_{14} - X_{23} = 0$, $X_{24} - X_{33} = 0$ in $\mathrm{Sym}(4,k)$ and so $H(4,k)$ is a linear variety of dimension 7.

Generalizing these examples, we have:

Proposition 20.37. $H(n,k)$ *is a linear variety of dimension* $n(n + 1)/2 - (n-1)(n-2)/2 = 2n - 1$ *and hence,* $\mathrm{Hank}(n,k)$ *is nonsingular.*

Proof: We use induction. Assume that $H(n,k)$ is a linear variety of dimension $2n - 1$. If $H = (h_{ij})_{i,j=1,\ldots,n+1} \in H(n+1,k)$, then $H_n = (h_{ij})_{i,j=1,\ldots,n} \in H(n,k)$ satisfies $(n-1)(n-2)/2$ linear equations in X_{ij} ($i \leq j$, $i,j = 1,2,\ldots,n$) and the elements h_{in+1} of the last column of H satisfy the $n - 1$ linear equations

$$X_{in+1} - X_{i+1n} = 0 \tag{20.38}$$

$i = 1, \ldots, n - 1$. Conversely, any solution of the $(n-1)(n-2)/2$ equations of $H(n,k)$ and the $n - 1$ equations (20.38) determines an element of $H(n+1,k)$. Since $(n-1) + (n-1)(n-2)/2 = n(n-1)/2$, $H(n+1,k)$ is a linear variety of dimension $[(n+1)(n+2) - n(n-1)]/2 = 2(n+1) - 1$ and the proposition is established.

Now let us give several more interpretations of the tangent space in terms of the local ring $\mathcal{O}_{\xi,V}$ of a point and the maximal ideal $\mathfrak{m}_\xi$ of a point. If $V \subset A_k^N$ is an affine variety, $\mathfrak{p} = I(V)$, and $\xi \in V$, then we recall (section 9) that $\mathcal{O}_{\xi,V} = \{f/g : f,g \in k[V], g(\xi) \neq 0\}$. It follows that if $D \in \mathrm{Der}_k(k[V], k_\xi)$, then D extends uniquely to a derivation D' in $\mathrm{Der}_k(\mathcal{O}_{\xi,V}, k_\xi)$ by setting $D'(f/g) = (g(\xi)Df - f(\xi)Dg)/g(\xi)^2$. In other words, $T_{V,\xi} = \mathrm{Der}_k(\mathcal{O}_{\xi,V}, k_\xi)$ so that the tangent space at ξ can be identified with the space of k-derivations of $\mathcal{O}_{\xi,V}$ which are centered at ξ. Let $R = k[V]$ and let $m_\xi = \{f \in R : f(\xi) = 0\}$ be the maximal ideal of regular functions on V which vanish at ξ. Then m_ξ/m_ξ^2 is a k $(= k_\xi)$ vector space. If $D \in T_{V,\xi} = \mathrm{Der}_k(R, k_\xi) = \mathrm{Der}_k(\mathcal{O}_{\xi,V}, k_\xi)$, then $D(m_\xi^2) = 0$ so that D defines a linear functional on m_ξ/m_ξ^2 i.e. D may be viewed as an element of the $(k-)$ dual space $(m_\xi/m_\xi^2)^*$. Conversely, if ℓ is a linear functional on m_ξ/m_ξ^2, then the map $D_\ell : R \to k_\xi$ given by

$$D_\ell(f) = \ell(\overline{f - f(\xi)}) \tag{20.39}$$

where $\overline{f - f(\xi)}$ is the m_ξ^2 residue of $f - f(\xi)$, is an element of $\mathrm{Der}_k(R, k_\xi)$ $= T_{V,\xi}$. Thus, we have:

Proposition 20.40. $T_{V,\xi} = (m_\xi/m_\xi^2)^*$.

Summarizing we have shown that $T_{V,\xi} = V(\Sigma \frac{\partial f}{\partial X_i}(\xi)(X_i - \xi_i)) = \mathrm{Der}_k(R, k_\xi) = \mathrm{Der}_k(\mathcal{O}_{\xi,V}, k_\xi) = (m_\xi/m_\xi^2)^*$.

Now we have established three of the major results of scalar linear system theory, namely: (i) that the Laurent map L is a k-isomorphism between $\mathrm{Rat}(n, k)$ and $\mathrm{Hank}(n, k)$ (Theorem 8.12 and Corollary 9.15); (ii) that there exist minimal realizations (Theorem 10.18) and that any two minimal realizations are isomorphic with a *unique isomorphism* (the state space isomorphism theorem 11.19); and, (iii) that $(\mathrm{Char}(n, k), \mathfrak{R}_\chi)$ is a geometric quotient of $S_{1,1}^n$ modulo $GL(n, k)$ where $\mathrm{Char}(n, k)$ is a nonsingular quasi-affine variety of dimension $2n$ (theorem 14.20 and corollary 20.24). The fourth major result is the so-called pole placement theorem which we prove in the next section.

21. Feedback and the Pole Placement Theorem

We now turn our attention to the study of feedback. Let $G = GL(n, k)$ and let $k^* = GL(1, k) = k - \{0\} = \mathbb{A}_k^1 - \{0\}$. We note that G and k^* are (quasi-affine) varieties. Now let K be an element of $(k^n)'$. Then K may be viewed as an element of $\mathrm{Hom}_k(k^n, k)$ which in turn may be viewed as $\mathbb{A}_k^n$ (written as a row). Consider the set $\Gamma_f(n, 1) = G \times \mathrm{Hom}_k(k^n, k) \times k^* = G \times \mathbb{A}_k^n \times k^*$ and *define* a multiplication in $\Gamma_f(n, 1)$ as follows:

$$[g, K, \alpha][g_1, K_1, \alpha_1] = [gg_1, Kg_1 + \alpha K_1, \alpha\alpha_1] \qquad (21.1)$$

where $g, g_1 \in G$, K, $K_1 \in \mathbb{A}_k^n$, and $\alpha, \alpha_1 \in k^*$. We then have:

Theorem 21.2. (i) $\Gamma_f(n, 1)$ *is a group and a quasi affine variety;* (ii) *the mappings* $\mu : \Gamma_f(n, 1) \times \Gamma_f(n, 1) \to \Gamma_f(n, 1)$ *and* $i : \Gamma_f(n, 1) \to \Gamma_f(n, 1)$ *given by* $\mu(\gamma, \gamma_1) = \gamma\gamma_1$ *and* $i(\gamma) = \gamma^{-1}$, *where* $\gamma, \gamma_1 \in \Gamma_f(n, 1)$, *are morphisms;* (iii) *the mapping* $\tau : \Gamma_f(n, 1) \to GL(n + 1, k)$ *given by*

$$\tau([g, K, \alpha]) = \begin{bmatrix} g & 0 \\ K & \alpha \end{bmatrix} \qquad (21.3)$$

is an isomorphism of $\Gamma_f(n, 1)$ *onto a closed subgroup* $\tilde{\Gamma}_f(n, 1)$ *of* $GL(n + 1, k)$; (iv) $\tilde{\Gamma}_f(n, 1)$ *is a linear subvariety of* $GL(n + 1, k)$ *(therefore, nonsingular) of dimension* $n^2 + n + 1 = (n + 1)^2 - n$; *and,* (v) $\Gamma_f(n, 1)$ *acts on* $\mathbb{A}_k^{n^2+2n}$ *via the morphism* $\phi : \Gamma_f \times \mathbb{A}_k^{n^2+2n} \to \mathbb{A}_k^{n^2+2n}$ *given by*

$$\phi([g, K, \alpha], (A, b, c)) = (g(A - b\alpha^{-1}K)g^{-1}, gb\alpha^{-1}, cg^{-1}) \qquad (21.4)$$

or, equivalently, $\tilde{\Gamma}_f(n, 1)$ *acts on* $\mathbb{A}_k^{n^2+2n}$ *via the morphism* $\tilde{\phi} : \tilde{\Gamma}_f(n, 1) \times \mathbb{A}_k^{n^2+2n} \to \mathbb{A}_k^{n^2+2n}$ *given by*

$$\tilde{\phi}\left(\begin{bmatrix} g & 0 \\ K & \alpha \end{bmatrix}, (A, b, c) \right) = \begin{bmatrix} g & 0 \\ 0 & 1 \end{bmatrix} \begin{bmatrix} A & b \\ c & 0 \end{bmatrix} \begin{bmatrix} g^{-1} & 0 \\ -\alpha^{-1}Kg^{-1} & \alpha^{-1} \end{bmatrix} \qquad (21.5)$$

for $g \in G$, $K \in \mathbb{A}_k^n$, *and* $\alpha \in k^*$.

Proof: Since $\Gamma_f(n,1) = G \times A_k^n \times k^*$, it is clearly a quasi affine variety. The identity in $\Gamma_f(n,1)$ is $[I,0,1]$ and if $\gamma = [g, K, \alpha]$, then $\gamma^{-1} = [g^{-1}, -\alpha^{-1}Kg^{-1}, \alpha^{-1}]$. Moreover, the multiplication (21.1) is clearly associative. Thus, (i) is established. As for (ii), both μ and i are given by regular functions and so, are morphisms. We note as well that τ is clearly a morphism. Since

$$\tau([I,0,1]) = \begin{bmatrix} I & 0 \\ 0 & 1 \end{bmatrix} \quad , \quad \tau([g,K,\alpha]^{-1}) = \left(\begin{bmatrix} g & 0 \\ K & \alpha \end{bmatrix} \right)^{-1}$$

and

$$\tau([g,K,\alpha][g_1,K_1,\alpha_1]) = \begin{bmatrix} gg_1 & 0 \\ Kg_1 + \alpha K_1 & \alpha\alpha_1 \end{bmatrix}$$
$$= \begin{bmatrix} g & 0 \\ K & \alpha \end{bmatrix} \begin{bmatrix} g_1 & 0 \\ K_1 & \alpha_1 \end{bmatrix}$$

for $g \in G$, $K \in A_k^n$, and $\alpha \in k^*$, τ is a group homomorphism. Moreover, τ is obviously injective. If $\tilde{\Gamma}_f(n,1) = \tau(\Gamma_f(n,1))$, then $\tilde{\Gamma}_f(n,1)$ is the linear subvariety of $GL(n+1,k)$ defined by the equations

$$X_{in+1} = 0 \quad , \quad i = 1, \ldots, n \tag{21.6}$$

and so, $\tilde{\Gamma}_f(n,1)$ is nonsingular (corollary 20.23) and has dimension $(n+1)^2 - n = n^2 + n + 1$. If

$$\tilde{g} = \begin{bmatrix} g & 0 \\ K & \alpha \end{bmatrix} \in \tilde{\Gamma}_f(n,1)$$

then $\tau^{-1}(\tilde{g}) = [g, K, \alpha]$ and so, τ^{-1} is also a morphism. In other words, (iii) and (iv) hold and τ is an isomorphism of algebraic groups.

If $\gamma = [g, K, \alpha]$ and $\gamma_1 = [g_1, K_1, \alpha_1]$, then

$$(\gamma\gamma_1) \cdot (A, b, c) = \begin{bmatrix} gg_1 & 0 \\ 0 & 1 \end{bmatrix} \begin{bmatrix} A & b \\ c & 0 \end{bmatrix}$$
$$\begin{bmatrix} (gg_1)^{-1} & 0 \\ -(\alpha\alpha_1)^{-1}(Kg_1 + \alpha K_1)(gg_1)^{-1} & (\alpha\alpha_1)^{-1} \end{bmatrix}$$
$$= \begin{bmatrix} g & 0 \\ 0 & 1 \end{bmatrix} \begin{bmatrix} g_1 & 0 \\ 0 & 1 \end{bmatrix} \begin{bmatrix} A & b \\ c & 0 \end{bmatrix}$$
$$\begin{bmatrix} g_1^{-1} & 0 \\ -\alpha_1^{-1}K_1g_1^{-1} & \alpha_1^{-1} \end{bmatrix} \begin{bmatrix} g^{-1} & 0 \\ -\alpha^{-1}Kg^{-1} & \alpha^{-1} \end{bmatrix}$$

and so, ϕ or $\tilde{\phi}$ define actions of $\Gamma_f(n,1)$ or $\tilde{\Gamma}_f(n,1)$ on $\mathbb{A}_k^{n^2+2n}$.

Since $\Gamma_f(n,1)$ and $\tilde{\Gamma}_f(n,1)$ are isomorphic, we shall not distinguish between them and shall use either form interchangeably. We then have:

Definition 21.7.   The group $\Gamma_f(n,1)$ is called the *state feedback group (of type $(n,1)$)* and the action of $\Gamma_f(n,1)$ on $\mathbb{A}_k^{n^2+2n}$ is called *state feedback.*

For simplicity, we write Γ_f in place of $\Gamma_f(n,1)$ in the remainder of this section. If $x = (A,b,c) \in \mathbb{A}_k^{n^2+2n}$ and if $\gamma \in \Gamma_f$, then $\phi(\gamma, x) = \gamma \cdot x \in \mathbb{A}_k^{n^2+2n}$ and ϕ is a morphism of $\Gamma_f \times \mathbb{A}_k^{n^2+2n}$ into $\mathbb{A}_k^{n^2+2n}$. We want to study the properties of this morphism and of state feedback.

Proposition 21.8. *Let* $\mathbf{Y}(A,b,c) = [b \quad Ab \quad \cdots \quad A^{n-1}b]$ *be the controllability map (section 11). Then the rank of* $\mathbf{Y}(A,b,c)$ *is invariant under state feedback and so, controllability is invariant under state feedback.*

Proof: Let $\gamma = [g, K, \alpha] \in \Gamma_f$ and let $x = (A,b,c)$, $x_1 = (A_1, b_1, c_1)$ $= \gamma \cdot x$ so that $A_1 = g(A - b\alpha^{-1}K)g^{-1}$, $b_1 = gb\alpha^{-1}$ and $c_1 = cg^{-1}$. Then

$$[b_1 \ A_1 b_1 \ \cdots \ A_1^{n-1} b_1] = g[b \ (A - b\alpha^{-1}K)b \ \cdots (A - b\alpha^{-1}K)^{n-1}b]\alpha^{-1}$$

and range $\mathbf{Y}(A_1, b_1, c_1) = g$ range $\mathbf{Y}(A,b,c)\alpha^{-1}$ so that

$$\text{rank } \mathbf{Y}(A_1, b_1, c_1) = \text{rank } \mathbf{Y}(A,b,c).$$

Example 21.9.   Let $A = \begin{bmatrix} 0 & 1 \\ 1 & 1 \end{bmatrix}$, $b = \begin{bmatrix} 1 \\ 0 \end{bmatrix}$, $c = [1 \quad 1]$. Then $\mathbf{Y}(A,b,c) = \begin{bmatrix} 1 & 0 \\ 0 & 1 \end{bmatrix}$, $\mathbf{Z}(A,b,c) = \begin{bmatrix} 1 & 1 \\ 1 & 2 \end{bmatrix}$ so that (A,b,c) is minimal. Let $\gamma = [I, (1,2), 1]$. Then

$$\gamma \cdot (A,b,c) = (A_1, b_1, c_1) = \left(\begin{bmatrix} -1 & -1 \\ 1 & 1 \end{bmatrix}, \begin{bmatrix} 1 \\ 0 \end{bmatrix}, [1 \quad 1] \right)$$

and $\mathbf{Z}(A_1, b_1, c_1) = \begin{bmatrix} 1 & 1 \\ 0 & 0 \end{bmatrix}$ so that (A_1, b_1, c_1) is *not* observable. In other words, observability (a fortiori minimality) is not invariant under state feedback.

Proposition 21.10.   *Let* $N_f = \{[I, K, 1] \in \Gamma_f\}$ *and let* $G_f = \{[g, 0, \alpha] \in \Gamma_f\}$. *Then* (i) N_f *is a normal subgroup of* Γ_f; (ii) G_f *is a subgroup of* Γ_f *which acts on* N_f *via inner automorphisms; and,* (iii) $\Gamma_f = N_f G_f = G_f N_f$.

Proof: Clearly N_f and G_f are subgroups of Γ_f. If $\gamma = [g, K, \alpha] \in \Gamma_f$, then $\gamma = [I, Kg^{-1}, 1][g, 0, \alpha] = [g, 0, \alpha][I, \alpha^{-1}K, 1]$ so that $\Gamma_f = N_f G_f = G_f N_f$. Since

$$[g, K, \alpha][I, K_1, 1][g^{-1}, -\alpha^{-1}Kg^{-1}, \alpha^{-1}] = [I, \alpha^{-1}K_1 g^{-1}, 1]$$

and

$$[g, 0, \alpha][I, K_1, 1][g^{-1}, 0, \alpha^{-1}] = [I, \alpha^{-1}K_1 g^{-1}, 1],$$

N_f is a normal subgroup on which G_f acts via inner automorphisms.

Now let $x = (A, b, c)$ and $f(z) = f_x(z) = c(zI - A)^{-1}b$. If $\gamma_g = [g, 0, \alpha] \in G_f$ and $x_g = \gamma_g \cdot x$, then $f_{x_g}(z) = (cg^{-1})(zI - gAg^{-1})^{-1}gb\alpha^{-1} = f_x(z)\alpha^{-1}$. It follows that if $\gamma \in G_f$, then the poles of $f_{\gamma \cdot x}(z)$ and $f_x(z)$ are the same and are determined by the roots of $\det(zI - A) = q_A(z)$. On the other hand, if $\gamma_K = [I, K, 1] \in N_f$ and $x_K = \gamma_K \cdot x$, then $f_{x_K}(z) = c(zI - A + bK)^{-1}b$ and the poles of $f_{x_K}(z)$ are determined by the roots of $\det(zI - A + bK) = q_{A-bK}(z)$. Thus, to analyze the effect of state feedback on the poles of $f_x(z)$ we need only study the action of N_f on the orbits under the action of G_f. Put in another way, if $\mathcal{P}_n(z) = \{q(z) : q(z) \text{ is a monic polynomial of degree } n\}$ and if we define a mapping $\Delta : \Gamma_f \times \mathbb{A}_k^{n^2+2n} \to \mathcal{P}_n(z)$ by setting

$$\Delta(\gamma, x) = \det(zI - g(A - b\alpha^{-1}K)g^{-1}) \qquad (21.11)$$

where $\gamma = [g, K, \alpha] \in \Gamma_f$, $x = (A, b, c) \in \mathbb{A}_k^{n^2+2n}$, then $\Delta(\gamma, x) = \Delta(\gamma_e, \gamma \cdot x)$ where $\gamma_e = [I, 0, 1]$ so that

$$\Delta(\gamma_1 \gamma_2, x) = \Delta(\gamma_1, \gamma_2 x) \qquad (21.12)$$

for all γ_1, γ_2 in Γ_f. If $\gamma_g \in G_f$, then $\Delta(\gamma_g, x) = \Delta(\gamma_e, \gamma_g x) = \det(zI - A) = \Delta(\gamma_e, x)$ so that Δ is constant on orbits under G_f (i.e. Δ is a G_f-invariant). It follows from (21.12) that the range $\Delta(\Gamma_f, x)$ is the same as the range $\Delta(N_f, x)$ and the range $\Delta(\Gamma_f, \gamma_g x)$ is the same as the range $\Delta(\Gamma_f, x)$ for $\gamma_g \in G_f$. Observing that $\mathcal{P}_n(z)$ can be identified with $\mathbb{A}_k^n$ by the mapping $a_0 + a_1 z + \cdots + a_{n-1} z^{n-1} + z^n \longrightarrow (a_0, a_1, \ldots, a_{n-1})$, we now have:

Problem 21.13 (The Coefficient Assignment Problem).   *Given* $x \in \mathbb{A}_k^{n^2+2n}$, *determine the range of the map* $\Delta_x : N_f \to \mathbb{A}_k^n$ *given by*

$$\Delta_x(K) = \Delta(\gamma_K, x) = (\chi_1(A - bK), \ldots, \chi_n(A - bK)) \qquad (21.14)$$

where the $\chi_i(\cdot)$ *are the characteristic coefficients, and, in particular, determine conditions for the surjectivity of* Δ_x.

If k has characteristic zero, then Δ_x will be surjective if and only if the map $T_x : N_f \to \mathbb{A}_k^n$ given by

$$T_x(K) = (\text{tr}(A + bK), \text{tr}(A + bK)^2, \ldots, \text{tr}(A + bK)^n) \qquad (21.15)$$

is surjective in view of the results in Appendix C (e.g. proposition C.14). The map T_x is called the *trace-assignment map* and we can formulate the following:

Problem 21.16 (The Trace Assignment Problem).   *Given* $x \in \mathbb{A}_k^{n^2+2n}$, *determine the range of* T_x *and, in particular, determine conditions for the surjectivity of* T_x.

We now turn our attention to solving these problems.

We first observe that since $zI - A + bK = (zI - A)[I + (zI - A)^{-1}bK]$, $\det[zI - A + bK] = \det[zI - A]\det[I + (zI - A)^{-1}bK]$. However,

$$\det[I + (zI - A)^{-1}bK] = 1 + K(zI - A)^{-1}b \qquad (21.17)$$

by a well-known determinant identity. It follows that

$$\Delta_x(K) = \Delta_x(0) + K\text{adj}(zI - A)b \qquad (21.18)$$

and hence that the range of $\Delta_x(K)$ is determined by the range of the *linear* map $\ell_x(K) : \mathbf{A}_k^n \to \mathbf{A}_k^n$ given by

$$\ell_x(K) = K \operatorname{adj}(zI - A)b \qquad (21.19)$$

where a polynomial of degree $n - 1$ is identified with its coefficients.

Lemma 21.20. $\ell_x(K) = K J_x$ *where*

$$J_x = [b \;\; Ab + \chi_1(A)b \;\; \cdots \;\; A^{n-1}b + \chi_1(A)A^{n-2}b + \cdots + \chi_{n-1}(A)b] \qquad (21.21)$$

and the χ_i are the characteristic coefficients.

Proof: An immediate consequence of the formula $\operatorname{adj}(zI - A) = \sum_{i=1}^n (z^{n-i} + \chi_1(A)z^{n-i-1} + \cdots + \chi_{n-i}(A))A^{i-1}$.

Corollary 21.22. $\ell_x(K) = K \mathbf{Y}(A, b, c) L_x$ *where*

$$L_x = \begin{bmatrix} 1 & \chi_1 & \chi_2 & \cdots & \chi_{n-1} \\ 0 & 1 & \chi_1 & \cdots & \chi_{n-2} \\ \vdots & \vdots & \vdots & & \vdots \\ 0 & 0 & 0 & \cdots & 1 \end{bmatrix} \qquad (21.23)$$

and so, $\dim \mathcal{R}(\ell_x) = \operatorname{rank} \mathbf{Y}(A, b, c)$.

Theorem 21.24 (Coefficient Assignment Theorem) ([W-2]). Δ_x *is surjective if and only if x is controllable.*

Proof: Δ_x is surjective if and only if ℓ_x is surjective and ℓ_x is surjective if and only if $\dim \mathcal{R}(\ell_x) = \operatorname{rank} \mathbf{Y}(A, b, c)$ is n.

We also note that corollary 21.22 characterizes the range of Δ_x.

Now let us examine the trace assignment problem. We begin with a lemma.

Lemma 21.25.   (i) $\operatorname{tr}(A + bK)^m = \sum_{j=0}^m \binom{m}{j} \operatorname{tr}(A^{m-j}(bK)^j)$; (ii) $(bK)^j = (bK)[\operatorname{tr}(bK)]^{j-1}$ for $j = 1, \ldots$; (iii) $\operatorname{tr}(A^{m-j}(bK)^j) = \operatorname{tr}(A^{m-j}bK)[\operatorname{tr}(bK)]^{j-1}$; *and,* (iv) $\operatorname{tr}(bK)^j = [\operatorname{tr}(bK)]^j$.

Proof: (i) follows from the linearity of the trace and the fact that $\operatorname{tr}(XY) = \operatorname{tr}(YX)$. (ii) follows by induction and the easy fact that $(bK)^2 = (bK) \cdot \operatorname{tr}(bK)$. (iii) follows from (ii) and the linearity of the trace. (iv) is also an immediate result of (ii).

We let $t_x(K)$ be the *linear* map given by

$$t_x(K) = (\operatorname{tr}(bK), \operatorname{tr}(AbK), \ldots, \operatorname{tr}(A^{n-1}bK)) \tag{21.26}$$

and let $t_x^i(K) = \operatorname{tr}(A^{i-1}bK)$, $i = 1, \ldots, n$ be the components of t_x. In view of lemma 21.25, we have

$$T_x(K) = T_x(0) + t_x(K)N_x(K) \tag{21.27}$$

where the $n \times n$ matrix $N_x(K)$ is given by

$$N_x(K) = \begin{bmatrix} 1 & 0 & \cdots & 0 \\ 0 & 2 & \cdots & 0 \\ \vdots & \vdots & & \vdots \\ 0 & 0 & \cdots & n \end{bmatrix} + \begin{bmatrix} 0 & t_x^1 & (t_x^1)^2 & \cdots & (t_x^1)^{n-1} \\ 0 & 0 & (3t_x^1) & \cdots & \vdots \\ \vdots & & & & \vdots \\ 0 & \cdots & \cdots & \cdots & 0 \end{bmatrix}$$

$$\tag{21.28}$$

i.e. $N_x(K)$ is the sum of a constant diagonal matrix and a matrix which depends only on $t_x^1(K)$ and has zero entries on and below the main diagonal.

Lemma 21.29. *Suppose that the characteristic of k is zero or is greater than n. Then T_x is surjective if and only if t_x is surjective.*

Proof: Clearly T_x is surjective if and only if $t_x N_x$ is surjective. If t_x is surjective and $(\alpha_1, \ldots, \alpha_n) \in \mathbb{A}_k^n$, then there is a K such that $t_x(K) = (\alpha_1, \ldots, \alpha_n)(N_x(K))^{-1}$ as $N_x(K)$ depends on α_1 alone. Conversely, if $t_x N_x$ is surjective and $(\beta_1, \ldots, \beta_n) \in \mathbb{A}_k^n$, then there is a K such that $t_x(K)N_x(K) = (\beta_1, \ldots, \beta_n)N_x(K)$ as $N_x(K)$ involves β_1 alone.

Theorem 21.30 (Trace Assignment Theorem). *Suppose that the characteristic of k is zero or is greater than n. Then T_x is surjective if and only if $x = (A, b, c)$ is controllable.*

Proof: In view of the previous lema, it is enough to show that t_x is surjective if and only if x is controllable. Since t_x is a linear map, t_x is surjective if and only if $J(t_x(K); K)$ is nonsingular. But $J(t_x(K); K) = [b \ Ab \cdots A^{n-1}b] = \mathbf{Y}(A, b, c)$ by lemma 21.31 and the result follows.

Lemma 21.31. *Let* $X \in M(n, n; k)$. *Then* $J(\mathrm{Tr}(XbK); K) = Xb$.

Proof: Simply note that $\mathrm{tr}(XbK) = (X^1 b)K_1 + \cdots + (X^n b)K_n$ where X^i is the i-th row of X.

Example 21.32.   Let $A = \begin{bmatrix} 0 & 1 \\ 0 & 0 \end{bmatrix}$, $b = \begin{bmatrix} 0 \\ 1 \end{bmatrix}$, $c = [1 \ \ 0]$ so that (A, b, c) is minimal. Suppose that k has characteristic 2. Then $\mathrm{tr}(A + bK) = K_2$ and $\mathrm{tr}(A + bK)^2 = (K_1 + K_1) + K_2^2 = K_2^2$ so that T_x is not surjective.

Example 21.33.   Let $A = \begin{bmatrix} 0 & 1 & 0 \\ 0 & 0 & 1 \\ 0 & 0 & 0 \end{bmatrix}$, $b = \begin{bmatrix} 0 \\ 0 \\ 1 \end{bmatrix}$, $c = [1 \ \ 0 \ \ 0]$ so that (A, b, c) is minimal. Then $\mathrm{tr}(A + bK) = K_3$, $\mathrm{tr}(A + bK)^2 = (K_2 + K_2) + K_3^2$, and $\mathrm{tr}(A + bK)^3 = (K_1 + K_1 + K_1) + (K_2 K_3 + K_2 K_3 + K_2 K_3) + K_3^3$. Thus T_x is not surjective if k has characteristic 2 or 3.

The theorems 2.14 and 21.30 provide the solution to the so-called "pole-placement problem".

We have thus established the four major algebro-geometric results of scalar linear system theory, namely: (i) that the Laurent map L is a k-isomorphism between $\mathrm{Rat}(n, k)$ and $\mathrm{Hank}(n, k)$; (ii) that minimal realizations exist and that any two minimal realizations are uniquely isomorphic; (iii) that $(\mathrm{Char}(n, k), \mathfrak{R}_x)$ is a geometric quotient of $S_{1,1}^n$ modulo $GL(n, k)$ with $\mathrm{Char}(n, k)$ a nonsingular quasi-affine variety of dimension $2n$; and, (iv) that the "poles" are assignable under state feedback if and only if the system is controllable.

In order to develop a similar theory for multivariable linear systems, we shall require ideas from projective algebraic geometry and ultimately, an even more general approach. With this in mind, we examine "abstract affine varieties" in the next section.

22. Affine Algebraic Geometry: Varieties

Let R be an affine k-algebra and let $X = \mathrm{Spm}(R) = \{\mathfrak{m} : \mathfrak{m}$ is a maximal ideal in $R\}$ be the maximal spectrum of R. We recall (theorem 7.5, corollary, 7.6) that $X = \mathrm{Spm}(R) \simeq \mathrm{Hom}_k(R, k)$ and that R may be viewed as the coordinate ring of some affine algebraic set. What we wish to do here, is to show that X may be made into a topological space, that the "regular" functions on X are simply the elements of R, and that if U is open in X, then the "regular" functions on U can be defined in a manner totally analogous to that used in section 9 (definition 9.8).

Proposition 22.1. *Let* R, S *be affine* k-*algebras and let* ψ *be an element of* $\mathrm{Hom}_k(R, S)$. *If* $\mathfrak{n} \in \mathrm{Spm}(S)$, *then* $\psi^{-1}(\mathfrak{n}) \in \mathrm{Spm}(R)$.

Proof: Let $\alpha_{\mathfrak{n}} \in \mathrm{Hom}_k(S, k)$ with $\mathrm{Ker}\,\alpha_{\mathfrak{n}} = \mathfrak{n}$. Then $\alpha_{\mathfrak{n}} \circ \psi \in \mathrm{Hom}_k(R, k)$ and $\mathrm{Ker}\,\alpha_{\mathfrak{n}} \circ \psi \in \mathrm{Spm}(R)$. Thus it will be enough to show that $\psi^{-1}(\mathfrak{n}) = \mathrm{Ker}(\alpha_{\mathfrak{n}} \circ \psi)$. If $r \in \psi^{-1}(\mathfrak{n})$, then $\psi(r) \in \mathfrak{n}$ and $r \in \mathrm{Ker}(\alpha_{\mathfrak{n}} \circ \psi)$. If $r \in \mathrm{Ker}(\alpha_{\mathfrak{n}} \circ \psi)$, then $\alpha_{\mathfrak{n}}(\psi(r)) = 0$ and $\psi(r) \in \mathrm{Ker}\,\alpha_{\mathfrak{n}} = \mathfrak{n}$.

We now make the following:

Definition 22.2. If E is a subset of R and $(E) = \mathfrak{a}$ is the ideal generated by E, then $V(E) = \{\mathfrak{m} \in X : \mathfrak{m} \supset E\}$ is the *closure of* E. If M is a subset of $X = \mathrm{Spm}(R)$, then $I(M) = \{r \in R : r \in \mathfrak{m}$ for all $\mathfrak{m}$ in $M\}$ is the *ideal of* M ($I(M)$ is clearly an ideal).

Proposition 22.3. (i) $V(E) = V(\mathfrak{a}) = V(\sqrt{\mathfrak{a}})$ *where* $\mathfrak{a} = (E)$; (ii) $V(0) = X$, $V(R) = \phi$; (iii) $V(\cup E_i) = \cap V(E_i)$; *and,* (iv) $V(\mathfrak{a} \cap \mathfrak{b}) = V(\mathfrak{a}\mathfrak{b}) = V(\mathfrak{a}) \cup V(\mathfrak{b})$

Proof: Clearly, $V(E) = V(\mathfrak{a}) \supseteq V(\sqrt{\mathfrak{a}})$. If $\mathfrak{m} \supset \mathfrak{a}$, then $\mathfrak{m} \supset \sqrt{\mathfrak{a}}$ as $f^n \in \mathfrak{a} \subset \mathfrak{m}$ implies $f \in \mathfrak{m}$. Thus, $V(\mathfrak{a}) = V(\sqrt{\mathfrak{a}})$. Parts (ii) and (iii) are obvious. If $\mathfrak{m} \in V(\mathfrak{a}) \cup V(\mathfrak{b})$, then $\mathfrak{m} \in V(\mathfrak{a} \cap \mathfrak{b}) \supseteq V(\mathfrak{a}\mathfrak{b})$. On the other hand, if $\mathfrak{m} \supset \mathfrak{a}\mathfrak{b}$ but $\mathfrak{m} \not\supset \mathfrak{a}$ and $\mathfrak{m} \not\supset \mathfrak{b}$, then there are $f \in \mathfrak{a}$,

$f \notin \mathfrak{m}$ and $g \in \mathfrak{b}, g \notin \mathfrak{m}$. But $fg \in \mathfrak{ab} \subset \mathfrak{m}$ gives a contradiction.

It is clear from the proposition that the sets $V(\mathfrak{a})$, if viewed as the closed sets, define a topology on $X = \mathrm{Spm}(R)$. This topology is called the **Zariski topology**. In view of the proposition, we have $I(V(\mathfrak{a})) = I(V(\sqrt{\mathfrak{a}}))$. Since $\sqrt{\mathfrak{a}} = \mathfrak{p}_1 \cap \ldots \cap \mathfrak{p}_r$ where the $\mathfrak{p}_i$ are prime ideals (corollary 6.11), we can see that $I(V(\mathfrak{a})) = \sqrt{\mathfrak{a}}$ and that $V(I(M)) = \overline{M}$ are immediate consequences of the following:

Lemma 22.4. *If $\mathfrak{p}$ is a prime ideal in the affine k-algebra R, then $\mathfrak{p} = \cap \mathfrak{m}, \mathfrak{m} \in \mathrm{Spm}(R), \mathfrak{p} \subset \mathfrak{m}$.*

Proof: By passage to $R/\mathfrak{p}$, we may suppose $\mathfrak{p} = (0)$. Since, in that case, $\mathfrak{p} = (0) \subset \mathfrak{m}$ for all $\mathfrak{m}$, we must show that if $r \neq 0$, then there is an $\mathfrak{m}$ with $r \notin \mathfrak{m}$ or, equivalently, there is an $\alpha \in \mathrm{Hom}_k(R, k)$ with $\alpha(r) \neq 0$. If $K = K(R)$ is the quotient field of R and $S = R[1/r]$ is an affine k-algebra (not the zero ring), then there is, by the Nullstellensatz, an $\alpha_s \in \mathrm{Hom}_k(S, k)$. But $1 = \alpha_s(1) = \alpha_s(r)\alpha_s(1/r)$ and so, if $\alpha = \alpha_s$ restricted to R, then $\alpha(r) \neq 0$.

Since $X = \mathrm{Spm}(R)$ is a topological space with the Zariski topology, we can seek to determine the irreducible subsets (cf. section 6) and to find a suitable base for the topology.

Definition 22.5. Let $X_f = X - V(f) = \{\mathfrak{m} : f \notin \mathfrak{m}\}$. Then X_f is a *principal open subset* of X (since $V(f)$ is closed, X_f is clearly open).

Proposition 22.6. *The X_f form a base for the Zariski topology on X and (i) $X_f \cap X_g = X_{fg}$ so that $X_{f^n} = X_f$; (ii) $X_f = \phi$ if and only if $f = 0$; (iii) $X_f = X$ if and only if f is a unit in R; (iv) $X_f = X_g$ if and only if $\sqrt{(f)} = \sqrt{(g)}$; and, (v) X is quasi-compact as is each X_f.*

Proof: If $V(\mathfrak{a})$ is closed with $\mathfrak{a} = \sqrt{\mathfrak{a}} \neq R$, then, for $\mathfrak{m} \notin V(\mathfrak{a})$, there is an f with $f \notin \mathfrak{m}$ (take $f \in \mathfrak{a}$) and so, $\mathfrak{m} \in X_f$. Thus, the X_f form a base. Since $V(fg) = V(f) \cup V(g)$, (i) is clear. (ii) is an obvious consequence of the proof of lemma 22.4 and (iii) is a result of the fact that every proper ideal is contained in a maximal ideal. As for (iv), observe that $V(f) = V(\sqrt{(f)}) = V(g) = V(\sqrt{(g)})$ if and only if $\sqrt{(f)} = \sqrt{(g)}$. To establish (v), note that $X = \cup X_{f_i}$ implies $(f_i) = R$

so that $1 = \sum_{j=1}^{n} g_j f_j$ (a finite sum) and hence, $X = \cup_{j=1}^{n} X_{f_j}$.

We recall that a topological space is *Noetherian* if any descending sequence of closed sets eventually stops i.e. if $V_1 \supset V_2 \supset \cdots$ then $V_n = V_{n_0}$ for $n \geq n_0$.

Proposition 22.7. (i) $\{\mathfrak{m}\}$ *is closed;* (ii) $X = \mathrm{Spm}(R)$ *is a Noetherian topological space; and,* (iii) $V(\mathfrak{a})$ *is irreducible if and only if $\mathfrak{a}$ is prime and so, X is irreducible if and only if R is an integral domain.*

Proof: Clearly, $I(V(\mathfrak{m})) = \mathfrak{m}$ so that $\{\mathfrak{m}\}$ is closed. If $V_1 = V(\mathfrak{a}_1) \supset V_2 = V(\mathfrak{a}_2) \supset \cdots$ is a descending sequence of closed sets, then $\mathfrak{a}_1 \subset \mathfrak{a}_2 \subset \cdots$ is an ascending chain of ideals in R. Since R is Noetherian, $\mathfrak{a}_n = \mathfrak{a}_{n_0}$ for $n \geq n_0$ and some n_0 so that $V(\mathfrak{a}_n) = V(\mathfrak{a}_{n_0})$. As for (iii), if $V(\mathfrak{a})$ is irreducible and $fg \in \mathfrak{a}$, then $V(\mathfrak{a}) \subset V(f) \cup V(g)$ implies (say) $V(\mathfrak{a}) \subset V(f)$ and $f \in \mathfrak{a}$. On the other hand if $\mathfrak{p}$ is prime and $V(\mathfrak{p}) = V(\mathfrak{a}_1) \cup V(\mathfrak{a}_2)$, then $\mathfrak{p} = \mathfrak{a}_1 \cap \mathfrak{a}_2$ implies $\mathfrak{p} = \mathfrak{a}_1$ (say) and $V(\mathfrak{p})$ is irreducible.

Corollary 22.8. *The (irreducible) components of X are the closed sets $V(\mathfrak{p}_j)$ where $(0) = \cap \mathfrak{p}_j$ and the $\mathfrak{p}_j$ are minimal prime ideals in R.*

Now suppose that R and S are affine k-algebras and that $X = \mathrm{Spm}(R)$, $Y = \mathrm{Spm}(S)$. If $\psi \in \mathrm{Hom}_k(R, S)$, then ψ induces a map $\tilde{\psi} : Y \to X$ given by

$$\tilde{\psi}(\mathfrak{n}) = \psi^{-1}(\mathfrak{n})$$

where $\mathfrak{n} \in \mathrm{Spm}(S)$, in view of proposition 22.1.

Proposition 22.9. (i) $\tilde{\psi}^{-1}(X_f) = Y_{\psi(f)}$; (ii) $\tilde{\psi}^{-1}(V(\mathfrak{a})) = V(\psi(\mathfrak{a}))$ *where $\mathfrak{a}$ is an ideal in R;* (iii) $\overline{\tilde{\psi}(V(\mathfrak{b}))} = V(\psi^{-1}(\mathfrak{b}))$ *where $\mathfrak{b}$ is an ideal in S;* (iv) $\tilde{\psi}(Y)$ *is dense in X if and only if ψ is injective (cf. proposition 7.18); and,* (v) *if ψ is surjective, then $Y \simeq V(\ker \psi)$ (cf. proposition 7.18).*

Proof: Properties (iv) and (v) have been established in section 7. If $\mathfrak{n} \in Y - V(\psi(f))$, then $\psi(f) \notin \mathfrak{n}$ and $f \notin \psi^{-1}(\mathfrak{n})$. On the other hand, if $\mathfrak{n} \in \tilde{\psi}^{-1}(X - V(f))$, then $\psi(f) \notin \mathfrak{n}$, and so, (i) is established. As for (ii), noting that $\psi(f) = f \circ \tilde{\psi}$, we see that $\mathfrak{n} \in \tilde{\psi}^{-1}(V(\mathfrak{a}))$ if and only if $\tilde{\psi}(\mathfrak{n}) \in V(\mathfrak{a})$ if and only if $\psi^{-1}(\mathfrak{n}) \supset \mathfrak{a}$ if and only if $\mathfrak{n} \supset \psi(\mathfrak{a})$. To demonstrate (iii), we need only show that $\psi^{-1}(\mathfrak{b}) = I(\tilde{\psi}(V(\mathfrak{b}))$ as $\psi^{-1}(\sqrt{\mathfrak{b}}) = \sqrt{\psi^{-1}(\mathfrak{b})}$. But $g \in I(\tilde{\psi}(V(\mathfrak{b}))$ implies $(g \circ \tilde{\psi})(\mathfrak{n}) = 0$ for $\mathfrak{n} \in V(\mathfrak{b})$. In other words, $\psi(g)(\mathfrak{n}) = 0$ and hence, $\psi(g) \in I(V(\mathfrak{b})) = \mathfrak{b}$ or $g \in \psi^{-1}(\mathfrak{b})$.

Corollary 22.10. *$\tilde{\psi}$ is continuous.*

Corollary 22.11. *If $X = \mathrm{Spm}(R)$ is irreducible and $f \in R$ and $\psi : R \to R_f$ is the canonical injection, then $\tilde{\psi}^{-1}(X_f) = \mathrm{Spm}(R_f)$ and there is a natural bijective homeomorphism of X_f onto $\mathrm{Spm}(R_f)$ (where X_f has the induced Zariski topology). In other words, $X_f = \mathrm{Spm}(R_f)$.*

Proof: Since $\tilde{\psi}^{-1}(X_f) = Y_{\psi(f)} = Y_f$ and f is a unit in R_f, we have $Y_f = Y = \mathrm{Spm}(R_f)$. The rest of the corollary has been established in propositions 8.1 and 9.14.

If $\mathfrak{m}$ is an element of $X = \mathrm{Spm}(R)$, then $R_{\mathfrak{m}}$ is a local ring and we have:

Proposition 22.12. *$R = \cap R_{\mathfrak{m}}$ where $\mathfrak{m} \in X = \mathrm{Spm}(R)$.*

Proof: Clearly, $R \subset \cap R_{\mathfrak{m}}$. If $h \in \cap R_{\mathfrak{m}}$ and $\mathfrak{a}_h = \{g \in R : gh \in R\}$, then $\mathfrak{a}_h$ is an ideal in R. If $\mathfrak{a}_h$ is a proper ideal, then (by the Nullstellensatz) $V(\mathfrak{a}_h) \neq \phi$ and there is an $\mathfrak{m} \in X$ with $\mathfrak{a}_h \subset \mathfrak{m}$. But $h \in R_{\mathfrak{m}}$ implies $h = h_1/h_2$ with $h_2 \notin \mathfrak{m}$. However, $h_2 h = h_1 \in R$ so that $h_2 \in \mathfrak{a}_h \subset \mathfrak{m}$. This is a contradiction and so $\mathfrak{a}_h = R$. In other words, $1 \in \mathfrak{a}_h$ and $h \in R$.

Corollary 22.13. *If X is irreducible, then $R_f = \cap R_{\mathfrak{m}}$ where $\mathfrak{m} \in X_f = \mathrm{Spm}(R_f)$.*

Proof: Simply apply the proposition and corollary 22.11.

Now let us write $x \in X$ in place of $\mathfrak{m} \in X$. We then have:

Definition 22.14. A function $f : X \to k$ is *regular at* x if $f \in R_x$. The set of regular functions at x is simply the local ring R_x which we denote by $\mathcal{O}_{x,X}$.

Proposition 22.15. *If f is regular at x, then there are $g, h \in R$ and a neighborhood V of x such that $h(y) \neq 0$ for $y \in V$ and $f(y) = g(y)/h(y)$ for $y \in V$. Conversely, if X is irreducible and there are $g, h \in R$ and a neighborhood V of x such that $h \neq 0$ on V and $f = g/h$ on V, then f is regular at x.*

Proof: For the first part, simply take $V = X_h$. For the second part, simply note that $hf = g$ on V which is dense in X so that $hf = g$ on X and $f = g/h \in R_x$.

Definition 22.16.   If U is open in X and $f : X \to k$, then f is *regular on U* if f is regular for each x in X. If X is irreducible, then all $R_x \subset K(R)$, the quotient field of R, and f is regular on U if and only if $f \in \cap_{x \in U} R_x$. We let $\mathcal{O}_X(U) = \cap_{x \in U} R_x$ and call $\mathcal{O}_X(U)$ the k-algebra of *regular functions on U*.

Proposition 22.17. *Suppose that X is irreducible. Then* (i) $\mathcal{O}_X(X_f) = R_f$ *and* $\mathcal{O}_X(X) = R$; (ii) *if* $V \subset U$ *are open sets and* $\rho_{U,V} : \mathcal{O}_X(U) \to \mathcal{O}_X(V)$ *is the restriction map i.e.* $\rho_{U,V}(f)(\mathfrak{m}) = f(\mathfrak{m})$ *for* $\mathfrak{m} \in V$ *and* $f \in \mathcal{O}_X(U)$, *then* $\rho_{U,V}$ *is an element of* $\mathrm{Hom}\,_k(\mathcal{O}_X(U), \mathcal{O}_X(V))$, $\rho_{U,U} = $ *identity, and if* $W \subset V \subset U$, *then* $\rho_{U,V} \circ \rho_{V,W} = \rho_{U,W}$; (iii) *if* $U = \cup U_i$, $f_1, f_2 \in \mathcal{O}_X(U)$ *and* $\rho_{U,U_i}(f_1) = \rho_{U,U_i}(f_2)$ *for all* i, *then* $f_1 = f_2$; *and,* (iv) *if* $U = \cup U_\alpha$ *and* $f_\alpha \in \mathcal{O}_X(U_\alpha)$ *with* $\rho_{U_\alpha, U_\alpha \cap U_\beta}(f_\alpha) = \rho_{U_\beta, U_\beta \cap U_\alpha}(f_\beta)$ *all* α, β, *then there is a (unique by* (iii)*)* f *in* $\mathcal{O}_X(U)$ *such that* $\rho_{U,U_\alpha}(f) = f_\alpha$.

Proof: (i) has been established in corollary 22.13, and (ii) is obvious. As for (iii), we may suppose that $\rho_{U,U_f}(f_1) = \rho_{U,U_f}(f_2)$ for all f. If $f_1 \neq f_2$, then $g = f_1 - f_2 \in \mathcal{O}_X(U)$ and there is an $x \in U_g$ (as $X_g \cap U \neq \phi$). But $\rho_{U,U_g}(f_1) = \rho_{U,U_g}(f_2)$ implies $f_1(x) = f_2(x)$, which is a contradiction. To prove (iv), we may assume that the U_α are principal open sets and so reduce to the following: let U_{f_i} cover U and let $s_i \in \mathcal{O}_U(U_{f_i})$ with $s_i = s_j$ on $U_{f_i} \cap U_{f_j}$, then there is an s in $\mathcal{O}_X(U)$ with $\rho_{U,U_{f_i}}(s) = s_i$. Since X is quasi-compact, we actually have $U = \cup_{i=1}^n U_{f_i}$ and $s_i = r_i/f_i^m$, $i = 1, \ldots, n$. Since $s_i = s_j$ on $U_{f_i} \cap U_{f_j} = U_{f_i f_j}$, we have $(r_i f_j^m)/(f_i f_j)^m = (r_j f_i^m)/(f_i f_j)^m$ in $U_{f_i f_j}$ and, therefore, $(r_i f_j^m - r_j f_i^m)(f_i f_j)^{m_{ij}} = 0$ in $\mathcal{O}_X(U)$. Since U is irreducible, $r_i f_j^m = r_j f_i^m$. Moreover, $U_{f_i} = U_{f_i^m}$ and so, $(f_1^m, \ldots, f_n^m) = \mathcal{O}_X(U)$. In other words, $1 = \sum_{i=1}^n g_i f_i^m$. Let $s = \sum_{i=1}^n g_i r_i$. Then

$s \in \mathcal{O}_X(U)$ and $sf_j^m = \Sigma g_i r_i f_j^m = \Sigma g_i f_i^m r_j = r_j$. Thus, $\rho_{U,U_{f_j}}(s) = r_j/f_j^m = s_j$ and we are done.

Now, what we have is a topological space X and, for each non-empty open subset U of X, a k-algebra $\mathcal{O}_X(U)$, and, a family of k-homomorphisms $\rho_{U,V} : \mathcal{O}_X(U) \to \mathcal{O}_X(V)$ for $U \supset V$, such that (ii), (iii), (iv) of proposition 22.17 are satisfied. The family $\mathcal{O}$ of $\mathcal{O}_X(U)$ and maps $\rho_{U,V}$ is called a *sheaf* (of functions) and the pair $(X, \mathcal{O})$ is called a *ringed space*. We then have:

Definition 22.18.   If R is an irreducible affine k-algebra, then the *ringed space* $(X, \mathcal{O}_X)$ where $X = \mathrm{Spm}(R)$ and $\mathcal{O}_X$ is the sheaf $\{\mathcal{O}_X(U), \rho_{U,V}\}$ is an *affine algebraic variety over* k.

If $(X, \mathcal{O}_X)$ and $(Y, \mathcal{O}_Y)$ are ringed spaces and $\psi : X \to Y$, then we say that ψ is a *morphism of ringed spaces* if (i) ψ is continuous, and (ii) for all open sets V in Y, $f \in \mathcal{O}_Y(V)$ implies $\psi^*(f) = f \circ \psi \in \mathcal{O}_X(\psi^{-1}(V))$. Thus, in this context, a *morphism of affine algebraic varieties* is simply a morphism of ringed spaces. We leave it to the reader to verify the consistency of these ideas with what we have done previously. This approach will turn out to be most fundamental.

23. Interlude

We seek, in this section to indicate, in a brief and heuristic fashion, some of the flavor of Part II: Multivariable Linear Systems and Projective Algebraic Geometry. While affine algebraic geometry is quite satisfactory and natural for scalar systems, the study of multi-input, multi-output systems requires projective geometric ideas. For simplicity, we shall assume that our algebraically closed field k is the complex numbers $\mathbb{C}$ throughout this section.

The classical mathematical reasons for introducing projective space are to complete intersections and to compactify. Suppose we consider the family of lines $y = mx + b$ in $\mathbb{A}^2_{\mathbb{C}}$. Given two distinct lines $y = m_1 x + b_1, y = m_2 x + b_2$, these two lines will intersect in a point if $m_1 \neq m_2$, but will not intersect if $m_1 = m_2$. In some sense, the lines $y = mx + b_1, y = mx + b_2$ meet at "infinity". In other words, it would be pleasing to assert that any two distinct lines meet in a point and not have to deal with the exceptional case of parallel lines. Similarly, if we consider a curve of degree 2, say: $y^2 = x^2 - 1$, then lines of the family $y = mx$ will meet the curve in two points unless $m = \pm 1$; in that case, there is no intersection but $(y/x)^2 \rightarrow 1$ as $x \rightarrow \infty$ (or $-\infty$) and so, the curve has the lines as asymptotes. Again, in some sense, the curve meets the asymptotic line at "infinity" and it would be desirable to avoid the exceptional case if possible. We can do this by extending the affine plane $\mathbb{A}^2_{\mathbb{C}}$ to the projective plane $\mathbb{P}^2_{\mathbb{C}}$ by adding "points at infinity".

More precisely, consider the set $\mathbb{A}^3_{\mathbb{C}} - \{(0,0,0)\}$ and define an equivalence relation on it via multiplication by elements of $\mathbb{C}^*$ i.e. $x = (x_0, x_1, x_2) \sim y = (y_0, y_1, y_2)$ if $y_i = \alpha x_i, \alpha \in \mathbb{C}^*$. The set of equivalence classes is the projective plane $\mathbb{P}^2_{\mathbb{C}}$ and is the same as the set of all 1-dimensional subspaces of $\mathbb{A}^3_{\mathbb{C}}$ (i.e. the set of all lines through the origin in $\mathbb{A}^3_{\mathbb{C}}$). Elements of $\mathbb{P}^2_{\mathbb{C}}$ are called points and any representative (ξ_0, ξ_1, ξ_2) of a point $\boldsymbol{\xi}$ is called a set of *homogeneous coordinates* of $\boldsymbol{\xi}$. Note that $(\xi_0, \xi_1, \xi_2) \neq (0,0,0)$ and that if, say $\xi_2 \neq 0$, then $(\xi_0/\xi_2, \xi_1/\xi_2, 1)$ is also a set of homogeneous coordinates of $\boldsymbol{\xi}$ and every point (x, y) in $\mathbb{A}^2_{\mathbb{C}}$ defines exactly one point with homogeneous coordinates $(x, y, 1)$ in $\mathbf{P}_{\mathbf{C}}^2$. The points $(\xi_0, \xi_1, 0)$ (note either $\xi_0 \neq 0$ or $\xi_1 \neq 0$) are called *points at infinity* and $H_\infty = \{(\xi_0, \xi_1, 0)\}$ (i.e. $\xi_2 = 0$) is called the *line at infinity*. A *line* in $\mathbf{P}_{\mathbf{C}}^2$ may be defined as the set of points which satisfy a homogeneous equation of the form

$$a_0 x_0 + a_1 x_1 + a_2 x_2 = 0 \qquad (23.1)$$

with $(a_0, a_1, a_2) \neq (0, 0, 0)$. Observe that if one set of homogeneous coordinates of a point $\boldsymbol{\xi}$ satisfies (23.1), then all sets do. The lines $y = mx + b$ in $\mathbf{A}_{\mathbf{C}}^2$ correspond to the lines $mx_0 - x_1 + bx_2 = 0$ in $\mathbf{P}_{\mathbf{C}}^2$ and then, the two distinct (parallel) lines $y = mx + b_1, y = mx + b_2$ correspond to the lines $mx_0 - x_1 + b_1 x_2 = 0, mx_0 - x_1 + b_2 x_2 = 0$ which intersect in the point $(1, m, 0)$ at infinity! Similarly, the curve $y^2 = x^2 - 1$ corresponds to the curve $x_1^2 = x_0^2 - x_2^2$ (homogeneous of degree 2) and meets the line $x_0 - x_1 = 0$ (corresponds to $y = x$) in the point $(1, 1, 0)$ and the line $x_0 + x_1 = 0$ (corresponds to $y = -x$) in the point $(1, -1, 0)$. We remark that the line $x_0 - x_1 = 0$ is tangent to the curve $x_0^2 - x_1^2 - x_2^2 = 0$ at $(1, 1, 0)$ and that the curve and line meet "twice" at this point. In other words, we are able to include all intersections by working in projective space.

Now consider $\mathbf{P}_{\mathbf{C}}^1$ as the set of 1-dimensional subspaces of $\mathbf{A}_{\mathbf{C}}^2 = \mathbf{C}^2$. We claim that $\mathbf{P}_{\mathbf{C}}^1$ is a one-point compactification of $\mathbf{C}$. A simple classical way to see this is to identify $\mathbf{P}_{\mathbf{C}}^1$ with the 2-sphere $S^2 = \mathbf{R}^2 \cup \{\infty\}$ which is the one-point compactification of $\mathbf{C}$. If we identify S^2 with the set $\{(u, v, w) : u^2 + v^2 + w^2 = 1\}$ in $\mathbf{R}^3$ and if we let $\mathbf{P}_{\mathbf{C}}^1 = (\mathbf{P}_{\mathbf{C}}^1 - \{\xi_\infty\}) \cup \{\xi_\infty\}$ where $\{\xi_\infty\} = (1, 0)$ then $\mathbf{P}_{\mathbf{C}}^1 - \{\xi_\infty\} = \{(\xi, 1) : \xi \in \mathbf{C}\} \simeq \mathbf{C}$ and the map $\psi_{\xi_\infty} : \mathbf{P}_{\mathbf{C}}^1 \to S^2$ given by

$$\psi_{\xi_\infty}(\xi_\infty) = (0, 0, 1)$$

$$\psi_{\xi_\infty}((\xi, 1)) = \left(\frac{2r \cos\theta}{r^2 + 1}, \frac{2r \sin\theta}{r^2 + 1}, \frac{r^2 - 1}{r^2 + 1} \right)$$

$$= \left(\frac{2a}{a^2 + b^2 + 1}, \frac{2b}{a^2 + b^2 + 1}, \frac{a^2 + b^2 - 1}{a^2 + b^2 + 1} \right) \qquad (23.2)$$

where $\xi = re^{i\theta} = r\cos\theta + ir\sin\theta = a + ib \in \mathbf{C}$, identifies $\mathbf{P}_{\mathbf{C}}^1$ and S^2. An alternative way with, as we shall see, more relevance to control involves subspaces. Let ξ be a point of $\mathbf{P}_{\mathbf{C}}^1$ and let V_ξ be ξ viewed as a 1-dimensional subspace of $\mathbf{C}^2$. Suppose W is any 1-dimensional subspace of C^2 with $V_\xi \oplus W = C^2$ and let $\sigma(W) = \{V : V$ a 1-dimensional subspace of C^2 and $\dim_C V \cap W \geq 1\}$. Note that here $\sigma(W) = \{W\}$ is a single point. If V_η is any element of P_C^1, then either $V_\eta \oplus W = C^2$ or $V_\eta \in \sigma(W)$ so that $V_\eta = W$. In case $V_\eta \oplus W = C^2$, there is a linear map $L : V_\xi \to W$ with V_η as graph. Since the dimension is 1, L is multiplication by an element of C. Thus, $P_C^1 - \{\xi\}$ is identified with C and $P_C^1 = (P_C^1 - \{\xi\}) \cup \{\xi\} = \{P_C^1 - \sigma(W)\} \cup \{\sigma(W)\} = C \cup \{\infty\}$ = one-point compactification of C.

Definition 23.3.   Let $C^{N+1} = \{(x_0, \ldots, x_N) : x_i \in C\}$ and call x, y in $C^{N+1} - \{0\}$ equivalent if $x = \lambda y, \lambda \in C^*(= C - \{0\})$. The set of equivalence classes P_C^N is called *projective N-space*.

Since the equivalence identifies points on the same line through the origin, P_C^N is the same as the set of 1-dimensional subspaces of A_C^{N+1}. A point $\xi \in P_C^N$ can be represented by an $N + 1$-tuple $(\xi_0, \ldots, \xi_N) \neq 0$ and another $N + 1$-tuple $(\xi_0', \ldots, \xi_N')$ defines the same point if and only if there is a λ in C^* such that $\xi_j = \lambda \xi_j', j = 0, \ldots, N$. Any $N + 1$-tuple representing ξ is called a set of *homogeneous coordinates of* ξ.

Definition 23.4.   A point $\xi \in P_C^N$ is a *zero* of a polynomial $F(X_0, \ldots, X_N)$ in $C[X_0, \ldots, X_N]$ if $F(\xi_0, \ldots, \xi_N) = 0$ for *all* sets of homogeneous coordinates of ξ.

We observe that if $F = F_0 + F_1 + \cdots + F_r, F_i$ a homogeneous polynomial of degree i, then $F(\lambda \xi_0, \ldots, \lambda \xi_N) = F_0(\xi_0, \ldots, \xi_N) + \lambda F_1(\xi_0, \ldots, \xi_N) + \cdots + \lambda^r F_r(\lambda_0, \ldots, \lambda_N)$ and so, $F(\lambda \xi_0, \ldots, \lambda \xi_N) = 0$ for all λ in C^* if and only if $F_i(\xi_0, \ldots, \xi_N) = 0, i = 0, \ldots, r$.

Definition 23.5. $V \subset P_C^N$ is a *projective algebraic set* if V is the set of common zeros of a family of homogeneous elements of $C[X_0, \ldots, X_N]$.

Viewing the projective algebraic sets as the closed sets, we define a topology, the *Zariski topology*, on P_C^N. We show, of course, that this is a topology in Part II.

Definition 23.6.   An irreducible (cf. Definition 6.1) projective algebraic set in P_C^N is a *projective variety* and an open subset of a variety is a *quasi-projective variety*.

We shall soon construct some examples with a control motivation. First, suppose that $f(z) = b_0 + \cdots + b_{n-1}z^{n-1}/a_0 + a_1 z + \cdots + a_{n-1}z^{n-1} + z^n = p(z)/q(z)$ with p, q relatively prime. Then we can define a map $\psi_f : C \to C \times C = C^2$ via

$$\psi_f(z) = (p(z), q(z)) \tag{23.7}$$

If we let $\xi_\infty = (0, 1)$ and $C = \{(1, \xi) : \xi \in C\}$ then we can "extend" ψ_f to a map, also denoted ψ_f of $P_C^1 \to P_C^1$ via

$$\psi_f((x_0, x_1)) = (b_0 x_0^n + \cdots + b_{n-1}x_0 x_1^{n-1}, a_0 x_0^n + \cdots + a_{n-1}x_0 x_1^{n-1} + x_1^n) \tag{23.8}$$

Noting that $\psi_f(\xi_\infty) = \xi_\infty$ and that $\psi_f((1, \xi)) = (p(\xi), q(\xi))$, we see that ψ_f does indeed map P_C^1 into P_C^1 as p, q are coprime. We also observe that the "poles" of $f(z)$ are the "points" $(1, \xi)$ such that $\psi_f((1, \xi)) = (1, 0)$. Now let us consider a system with $m = 2$ inputs and $p = 2$ outputs. Such a system can be represented by a transfer matrix $F(z)$ which is 2×2. If each entry $F_{ij}(z)$ of $F(z)$ is a strictly proper rational function, then $F(z)$ is a strictly proper rational transfer matrix. In such a case, we can write

$$F(z) = P(z)Q^{-1}(z) = P(z)\text{Adj}Q(z)/\det Q(z) \tag{23.9}$$

where $P(z), Q(z)$ are 2×2 matrices with polynomial entries and are "coprime" in the sense that there are polynomial matrices $X(z), Y(z)$ such that

$$X(z)P(z) + Y(z)Q(z) \equiv I_2 \tag{23.10}$$

(this generalizes the relation $a(z)p(z) + b(z)q(z) = 1$ of the scalar case). We can then define a map $\psi_F : C \to M(4, 2)$ of C into the set of 4×2 matrices (with entries in C) by

$$\psi_F(z) = \begin{bmatrix} P(z) \\ Q(z) \end{bmatrix} \tag{23.11}$$

and coprimeness means that the 4×2 matrix $\psi_F(z)$ has rank 2 for all z i.e. that the columns of $\psi_F(z)$ define a 2-dimensional subspace of C^4.

Let us denote the set of all 2-dimensional subspaces of $\mathbf{C}^4$ by $\mathrm{Gr}(2,4)$ (Gr for Grassmannian). By analogy, $\mathrm{Gr}(1,4) = \mathbf{P}_{\mathbf{C}}^3$. If we fix a basis in $\mathbf{C}^4$, then every element of $\mathrm{Gr}(2,4)$ determines an element of $M_*(4,2)$, the set of 4×2 matrices of rank 2, and conversely. Let v_0, v_1, v_2, v_3 denote the rows of an element of $M_*(4,2)$ (the v_i are elements of $\mathbf{C}^2$) and let $[v,w]$ denote the determinant of the matrix $\begin{bmatrix} v \\ w \end{bmatrix}$. Consider the map $\mathcal{P} : M_*(4,2) \to \mathbf{P}_{\mathbf{C}}^5$ given by

$$\mathcal{P}\left(\begin{bmatrix} v_0 \\ v_1 \\ v_2 \\ v_3 \end{bmatrix}\right) = ([v_0,v_1],[v_0,v_2],[v_0,v_3],[v_1,v_2],[v_1,v_3],[v_2,v_3]) \quad (23.12)$$

so that $p_0 = [v_0,v_1], p_1 = [v_0,v_2], \ldots, p_5 = [v_2,v_3]$. Since the matrices are of rank 2, at least one $p_i \neq 0$. Let $G = GL(2,\mathbf{C})$. Then G acts on $M_*(4,2)$ by multiplication on the right and defines an equivalence relation on $M_*(4,2)$ by

$$\begin{bmatrix} v_0 \\ v_1 \\ v_2 \\ v_3 \end{bmatrix} E_G \begin{bmatrix} w_0 \\ w_1 \\ w_2 \\ w_3 \end{bmatrix} \quad \text{if } v_i = w_i g \quad (23.13)$$

for $i = 0, \ldots, 3$ and some $g \in G$. Clearly, $M_*(4,2)/G$ "$=$" $\mathrm{Gr}(2,4)$ and $\mathcal{P}$ is an invariant for the equivalence. Thus, $\mathcal{P}(M_*(4,2))$ or better $\mathcal{P}(M_*(4,2)/G)$ may be identified with $\mathrm{Gr}(2,4)$. Now consider the homogeneous polynomial equation

$$p_0 p_5 + p_2 p_3 - p_4 p_1 = 0 \quad (23.14)$$

which defines a projective variety in $\mathbf{P}_{\mathbf{C}}^5$ which is of dimension 4. We claim that $\mathrm{Gr}(2,4)$ is this variety. Since $[v_0,v_1][v_2,v_3]+[v_0,v_3][v_1,v_2]-[v_1,v_3][v_0,v_2] = 0$ holds for any 2-vectors v_0,v_1,v_2,v_3 as a determinantal identity, $\mathrm{Gr}(2,4)$ is contained in the variety. In fact, they are equal. For example, suppose that $\boldsymbol{\xi} = (1,\xi_1,\ldots,\xi_5)$ satisfies (23.14), then $v_0 = (1,0), v_1 = (0 \ 1), v_2 = (-\xi_3 \ \xi_1), v_3 = (-\xi_4 \ \xi_2)$ gives an element of $M_*(4,2)$ with image $\boldsymbol{\xi}$. We can thus consider the map ψ_F (abuse of notation) as a map $\psi_F : \mathbf{P}_{\mathbf{C}}^1 \to \mathrm{Gr}(2,4) \subset \mathbf{P}_{\mathbf{C}}^5$.

We can also see that it is reasonable to view $(\mathrm{Gr}(2,4), \mathcal{P})$ as a "geometric quotient" (in a projective sense) of $M_*(4,2)$ modulo the action of G. If $\mathbf{M}$ is an element of $M_*(4,2)$ and $\mathcal{O}_G(\mathbf{M})$ is the orbit under G, then $\dim \mathcal{O}_G(\mathbf{M}) = \dim G = 4$ (since $S_G(\mathbf{M}) = \{I\}$) and so orbits are closed (cf. lemma 15.9). If we let $\mathbf{G}(p_0, p_1, \ldots, p_5) = p_0 p_5 + p_2 p_3 - p_4 p_1$, then $\mathbf{G}$ is a form of degree 2 and $\mathrm{Gr}(2,4) = \{\boldsymbol{\xi} = (\xi_0, \ldots, \xi_5) : \mathbf{G}(\boldsymbol{\xi}) = 0\}$. Since $\partial \mathbf{G}/\partial p_j$ are not all 0 at any point of $\mathrm{Gr}(2,4)$, the variety $\mathrm{Gr}(2,4)$ is non-singular (cf. section 20.) and hence, normal (shown in Part II). Thus the map $\mathcal{P}$ is an open map (cf. proposition 18.22). As for making the ring of invariants part of a geometric quotient plausible, we deal locally with the open set $\mathrm{Gr}(2,4)_0 = \{\boldsymbol{\xi} \in \mathrm{Gr}(2,4) : \xi_0 \neq 0\}$. This set can be identified with the hypersurface $x_5 + x_2 x_3 - x_4 x_1 = 0$ in $\mathbb{A}_{\mathbb{C}}^5$ (let $\xi_0 = 1$ and $x_i = \xi_i/\xi_0$) which has the coordinate ring $\mathbb{C}[x_1, \ldots, x_5]/(x_5 + x_2 x_3 - x_4 x_1)$. It is easy to see that $\mathcal{P}^{-1}(\mathrm{Gr}(2,4)_0)$ "=" $G \times \mathbf{V}$ where

$$\mathbf{V} = \left\{ \mathbf{M} \in M_*(4,2) : \mathbf{M} = \begin{bmatrix} I \\ v_2 \\ v_3 \end{bmatrix} \right\}$$

and that there is an action of G on $G \times \mathbf{V}$ given by $g(g_1, \mathbf{M}) = (g g_1, \mathbf{M})$ (cf. (19.49)). Moreover, $\mathbb{C}[G \times \mathbf{V}]^G = \mathbb{C}[\mathbf{V}]$ and so, as in section 19., we want $\mathcal{P}^*(\mathbb{C}[\mathrm{Gr}(2,4)_0])$ "=" $\mathbb{C}[\mathbf{V}]$. But if,

$$\mathbf{M} = \begin{bmatrix} 1 & 0 \\ 0 & 1 \\ v_{21} & v_{22} \\ v_{31} & v_{32} \end{bmatrix}$$

is in $\mathbf{V}$, then $\mathcal{P}(\mathbf{M}) = (1, v_{22}, v_{32}, -v_{21}, -v_{31}, v_{21} v_{32} - v_{22} v_{31})$ and so, letting $x_1 = v_{22}, x_2 = v_{32}, x_3 = -v_{21}, x_4 = -v_{31}, x_5 = v_{21} v_{32} - v_{22} v_{31}$, we see that our assertion about the ring of invariants is plausible. *

Example 23.15.   Let

$$P(z) = \begin{bmatrix} p_{11} & p_{12} \\ p_{21} & p_{22} \end{bmatrix}, Q(z) = \begin{bmatrix} z + q_{11} & q_{12} \\ z + q_{21} & z + q_{22} \end{bmatrix}$$

(with $\det P \neq 0$). Set $v_0(x_0, x_1) = (p_{11} x_0 \ \ p_{12} x_0), v_1(x_0, x_1) = (p_{21} x_0 \ \ p_{22} x_0), v_2(x_0, x_1) = (x_1 + q_{11} x_0 \ \ q_{12} x_0), v_3(x_0, x_1) = (x_1 + q_{21} x_0 \ \ x_1 + q_{22} x_0)$. Then

$$\psi_F(x_0, x_1) = ([v_0(x_0, x_1), v_1(x_0, x_1)], \ldots, [v_2(x_0, x_1), v_3(x_0, x_1)])$$

---

    *   Of course, these issues are dealt with in Part II.

and so, for example, $p_0(x_0, x_1) = x_0^2 \det P, p_3(x_0, x_1) = x_0^2(p_{21}q_{12} - q_{11}p_{22}) - p_{22}x_0x_1, p_5(x_0, x_1) = x_1^2 + (q_{11} - q_{12})x_0x_1 + x_0^2(q_{11}q_{22} - q_{12}q_{21})$. Note that $\psi_F(0, 1) = (0, \ldots, 0, 1)$ and that the "poles" of ψ_F are the points where the "curve" $\psi_F(\mathbb{P}_{\mathbb{C}}^1)$ meets the hyperplane $p_5 = 0$!

Now suppose that Z_α is an element of $\text{Gr}(2, 4)$ and let $\sigma(Z_\alpha) = \{V \in \text{Gr}(2, 4) : \dim_{\mathbb{C}}(V \cap Z_\alpha) \geq 1\}$ (note that either $V \oplus Z_\alpha = \mathbb{C}^4$ or $V \in \sigma(Z_\alpha)$). Let $\boldsymbol{\xi}_\alpha = (p_0(Z_\alpha), \ldots, p_5(Z_\alpha))$ and consider the homogeneous equation of degree 1

$$p_0(Z_\alpha)p_5 - p_1(Z_\alpha)p_4 + p_2(Z_\alpha)p_3 + p_3(Z_\alpha)p_2 - p_4(Z_\alpha)p_1 + p_5(Z_\alpha)p_0 = 0$$
$$(23.16)$$

in $\mathbb{P}_{\mathbb{C}}^5$. Let H_α be the projective variety defined by (23.16). Then $\sigma(Z_\alpha) = H_\alpha \cap \text{Gr}(2, 4)$ for if Z_α corresponds to $v_{\alpha,0}, v_{\alpha,1}, v_{\alpha,2}, v_{\alpha,3}$ and V corresponds to v_0, v_1, v_2, v_3, then $V \in \sigma(Z_\alpha)$ if and only if

$$\det \begin{bmatrix} v_{\alpha,0} & v_0 \\ v_{\alpha,1} & v_1 \\ v_{\alpha,2} & v_2 \\ v_{\alpha,3} & v_3 \end{bmatrix} = 0$$

(note this is a 4×4 matrix and expansion into determinants of 2×2 matrices shows that (23.16) holds). The variety $\sigma(Z_\alpha)$ is called a Schubert variety.

Now suppose that we have a system represented by a 2×2 transfer matrix $F(z)$ with associated map $\psi_F : \mathbb{P}_{\mathbb{C}}^1 \to \text{Gr}(2, 4)$ given by

$$\psi_F(x_0, x_1) = \boldsymbol{\mathcal{P}} \begin{pmatrix} P(x_1/x_0) \\ Q(x_1/x_0) \end{pmatrix} \tag{23.17}$$

suitably "homogenized". The roots of $\det Q(z)$ are the "poles" of F and we wish to examine the effects of output feedback on the "poles". In other words, given $K \in M(2, 2)$, what happens when Q is shifted to $Q - KP$ i.e. what are the roots of $\det(Q - KP)$? We observe that

$$\begin{bmatrix} I & P \\ K & Q \end{bmatrix} = \begin{bmatrix} I & 0 \\ K & I \end{bmatrix} \begin{bmatrix} I & P \\ 0 & Q - KP \end{bmatrix}$$

and hence that

$$\det(Q - KP) = \det\left(\begin{bmatrix} I & P \\ K & Q \end{bmatrix}\right)$$

Thus, if $(1, \xi)$ is to be a "closed loop pole" after "output feedback with gain K", then $\det(Q(\xi) - KP(\xi)) = 0$ and the 2-dimensional subspace

$$W_K = \text{span} \begin{bmatrix} I \\ K \end{bmatrix}$$

in $\mathbb{C}^4$ satisfies the condition $\dim(W_K \cap Z_\xi) \geq 1$ where Z_ξ is the 2-dimensional subspace

$$Z_\xi = \text{span} \begin{bmatrix} P(\xi) \\ Q(\xi) \end{bmatrix}$$

i.e. $W_K \in \sigma(Z_\xi)$. If we are given $n = \text{degree}(\det Q)$ points $\xi_1, \ldots, \xi_n$, then the output feedback problem becomes: does there exist a K such that $W_K \in \cap_{i=1}^n \sigma(Z_{\xi_i})$ or, in view of (23.16), does there exist a K such that $W_K \in \text{Gr}(2,4) \cap \cap_{i=1}^n H_{\xi_i}$? If say, $n \leq 4$, then "in general", $\dim \cap_{i=1}^n H_{\xi_i} \geq 4 - n$ and $\dim \text{Gr}(2,4) \cap \cap_{i=1}^n H_{\xi_i} \geq 0$ so that such a W_K exists. Observe that if $\xi_\infty = (0,1)$, then

$$Z_\infty = \text{span} \begin{bmatrix} 0 \\ I \end{bmatrix}$$

and so, $W_K \notin \sigma(Z_\infty)$ for all K. Thus, a "non-degeneracy" condition e.g. $\cap_{i=1}^n \sigma(Z_{\xi_i}) \cap \sigma(Z_\infty)) = \emptyset$ is required. This is a very brief indication by simple example of some of the relevance of intersections, Grassmannians and Schubert varieties to a control problem.*

Grassmannians and other projective ideas also impact multivariable control systems in the state-space and Hankel matrix representations. So let us again consider a system with $m = 2$ inputs, $p = 2$ outputs, and of "degree" $n = 2$. This system is described by a triple (A, B, C) where A is $n \times n$, B is $n \times m$ and C is $p \times n$. (We are, for simplicity, assuming all are 2×2.) The transfer matrix associated with (A, B, C) is given by

$$F(z) = C(zI - A)^{-1}B \tag{23.18}$$

---

* For much more detail, see the survey article: "Pole Assignment by Output Feedback" by C. I. Byrnes.

and is, of course, rational. We may define *controllability, observability,* and *Hankel* maps as follows:

$$\mathbf{Y}(A,B,C) = [B \quad AB \cdots A^{n-1}B] \tag{23.19}$$

$$\mathbf{Z}(A,B,C) = \begin{bmatrix} C \\ CA \\ \vdots \\ CA^{n-1} \end{bmatrix} \tag{23.20}$$

$$\mathbf{H}(A,B,C) = \mathbf{Z}(A,B,C)\mathbf{Y}(A,B,C) \tag{23.21}$$

so that $\mathbf{Y}$ is $n \times nm$, $\mathbf{Z}$ is $np \times n$, and $\mathbf{H}$ is $np \times nm$. So, in our sample case, $\mathbf{Y}$ is 2×4, $\mathbf{Z}$ is 4×2 and $\mathbf{H}$ is 4×4. If $B = [b_1\ b_2]$ and if we let $v_0 = b_1, v_1 = b_2, v_2 = Ab_1, v_3 = Ab_2$, and if we say that (A,B,C) is *controllable* if $\text{rank}[v_0\ v_1\ v_2\ v_3] = \text{rank}[B\ AB] = \text{rank}\,\mathbf{Y}(A,B,C) = 2$, then (A,B,C) is controllable means that $\mathbf{Y}(A,B,C)$ is an element of $M_*(2,4)$. We can embed $M_*(2,4)$ in $\text{Gr}(2,4) \subset \mathbb{P}_{\mathbb{C}}^5$ via the map (again denoted $\mathcal{P}$)

$$\mathcal{P}([v_0,v_1,v_2,v_3]) = ([v_0\ v_1],[v_0\ v_2],[v_0\ v_3],[v_1\ v_3],[v_2\ v_3]) \tag{23.22}$$

and we have a natural map $\mathcal{P}_Y$ from the set of controllable systems into $\text{Gr}(2,4)$ given by $\mathcal{P}_Y = \mathcal{P} \circ \mathbf{Y}$. Similarly, there is a map $\mathcal{P}_Z$ from the set of observable systems into $\text{Gr}(2,4)$ given by $\mathcal{P} \circ \mathbf{Z}$. Viewing the systems (A,B,C) as elements of $\mathbb{A}_{\mathbb{C}}^{12}$, we can see that $(\mathbb{A}_{\mathbb{C}}^{12})_c = \{(A,B,C) : (A,B,C) \text{ is controllable }\}$ is an open set (hence, a quasi-affine variety) since $(\mathbb{A}_{\mathbb{C}}^{12})_c = \cup_{i=0}^4 (\mathbb{A}_{\mathbb{C}}^{12})_{p_i}$ where $p_i(A,B,C) = p_i(A,B)$ with $p_0(A,B) = \det[b_1\ b_2], p_1(A,B) = \det[b_1\ Ab_1], p_2(A,B) = \det[b_1\ Ab_2], p_3(A,B) = \det[b_2\ Ab_1], p_4(A,B) = \det[b_2\ Ab_2]$ (note that $p_5(A,B) = \det[Ab_1\ Ab_2] = \det A \cdot p_0(A,B)$). The map $\mathbf{Y}$ is a "morphism" of the quasi-projective (no mistake as every affine or quasi-affine variety is quasi-projective - Part II) variety $(\mathbb{A}_{\mathbb{C}}^{12})_c$ into the projective variety $\text{Gr}(2,4)$. In fact, $\mathbf{Y}((\mathbb{A}_{\mathbb{C}}^{12})_c) \subseteq \cup_{i=0}^4 \text{Gr}(2,4)_{p_i}$ and does not contain the point $(0,0,\ldots,0,1)$ which corresponds to the subspace $[0\ 0\ I]$ (in fact, no point of the form $(0,\xi_1,\ldots,\xi_4,1)$). Of course, things are similar for observable systems.

Calling a system *minimal* if it is both controllable and observable and letting $S_{2,2}^2(= S_{p,m}^n)$ be the set of minimal systems, we have a map $\mathcal{P}_{Y,Z} : S_{2,2}^2 \to \text{Gr}(2,4) \times \text{Gr}(2,4) \subset \mathbb{P}_{\mathbb{C}}^5 \times \mathbb{P}_{\mathbb{C}}^5$ given by

$$\mathcal{P}_{Y,Z}(A,B,C) = (\mathcal{P}_Z(A,B,C), \mathcal{P}_Y(A,B,C)) \tag{23.23}$$

(the projective notion of a product is somewhat more complicated than its affine counterpart and will be treated in Part II). We note that $\dim S_{2,2}^2 = 12$ and that $\dim \text{Gr}(2,4) \times \text{Gr}(2,4) = 4 + 4 = 8$. Let $G = GL(\mathbb{C}, n) = GL(\mathbb{C}, 2)$ be the group of "coordinate changes" in state-space. Then G acts on $S_{2,2}^2$ via the map $\gamma : G \times S_{2,2}^2 \to S_{2,2}^2$ given by

$$\gamma((g, (A, B, C))) = (gAg^{-1}, gB, Cg^{-1}) \tag{23.24}$$

and this action is compatible with the action of G on $M_*(4, 2)$ and $M_*(2, 4)$. In other words, $\mathbf{Y}(g \cdot (A, B, C)) = g\mathbf{Y}(A, B, C)$, etc. Thus, $\mathcal{P}_{Y,Z}$ is an invariant and may be viewed as a map of $S_{2,2}^2/G \to M_*(4, 2)/G \times M_*(2, 4)/G$.

Now the transfer matrix $F(z)$ of (23.18) has an expansion

$$F(z) = \sum_{j=1}^{\infty} H_j z^{-j}, \ H_j = CA^{j-1}B \tag{23.25}$$

and an associated "block" Hankel matrix

$$\mathcal{H} = \begin{bmatrix} H_1 & H_2 & H_3 & \cdots \\ H_2 & H_3 & \cdots & \cdots \\ H_3 & & \vdots & \\ \vdots & & & \end{bmatrix} = (H_{i+j-1})_{i,j=1}^{\infty} \tag{23.26}$$

which is of rank 2 (when $H_j = CA^{j-1}B$). In view of the Cayley-Hamilton theorem, the matrix

$$\mathcal{H}_2^2 = \begin{bmatrix} H_1 & H_2 \\ H_2 & H_3 \end{bmatrix} = \begin{bmatrix} CB & CAB \\ CAB & CA^2B \end{bmatrix} \tag{23.27}$$

must also be of rank 2. In other words, if $(A, B, C) \in S_{2,2}^2$, then $\mathbf{H}(A, B, C) = \mathcal{H}_2^2(A, B, C)$ is of rank 2. Moreover, $\mathbf{H}(g \cdot (A, B, C)) = \mathbf{H}(A, B, C)$ so that $\mathbf{H}$ is an invariant for the action of G. Now, suppose, on the other hand, that $\mathcal{H} = (H_{i+j-1})_{i,j=1}^{\infty}$ is a "block" Hankel matrix of rank 2. Then we have:

Query 23.28.   Does there exist an (A, B, C) in $S_{2,2}^2$ such that $H_j = CA^{j-1}B$?

Let $\mathcal{H}^{\alpha_1}, \mathcal{H}^{\alpha_2}(\alpha_1 < \alpha_2)$ denote the first two independent rows of $\mathcal{H}$. Note that $1 \leq \alpha_1 \leq 2$ (otherwise $\mathcal{H} = 0$). Then there is a $2 \times 2\ C$ with

$$C \begin{bmatrix} \mathcal{H}^{\alpha_1} \\ \mathcal{H}^{\alpha_2} \end{bmatrix} = \begin{bmatrix} \mathcal{H}^1 \\ \mathcal{H}^2 \end{bmatrix} = [H_1\ H_2 \cdots]$$

and since $\mathcal{H}^{i+2}$ is a subrow of $\mathcal{H}^i$ for any i, there is a (unique) $2 \times 2\ A$ with

$$A \begin{bmatrix} \mathcal{H}^{\alpha_1} \\ \mathcal{H}^{\alpha_2} \end{bmatrix} = \begin{bmatrix} \mathcal{H}^{\alpha_1+2} \\ \mathcal{H}^{\alpha_2+2} \end{bmatrix}$$

If we let B be the 2×2 matrix consisting of the first two columns of $\mathcal{H}^\alpha$ where

$$\mathcal{H}^\alpha = \begin{bmatrix} \mathcal{H}^{\alpha_1} \\ \mathcal{H}^{\alpha_2} \end{bmatrix}$$

i.e. if $B = [\mathcal{H}^\alpha]_{j=1,2}$, then it follows that $CB = H_1, CAB = H_2, \ldots,$ $CA^{j-1}B = H_j, \ldots$, and that $\mathcal{H}_2^2 = \mathbf{Z}(A,B,C)\mathbf{Y}(A,B,C) = \mathbf{H}(A,B,C)$. We know that the rank of $\mathcal{H}_2^2, \rho(\mathcal{H}_2^2) \leq 2$ and we shall show that, in fact, $\rho(\mathcal{H}_2^2) = 2$. This will insure that $(A,B,C) \in S_{2,2}^2$. Let

$$\mathcal{H}_\ell^\ell = \begin{bmatrix} H_1 & H_2 & \cdots & H_\ell \\ H_2 & H_3 & \cdots & H_{\ell+1} \\ \vdots & \vdots & \vdots & \\ H_\ell & H_{\ell+1} & \cdots & H_{2\ell-1} \end{bmatrix}$$

and let λ be the least integer ℓ such that $\rho(\mathcal{H}_\ell^\ell) \leq 2$.

Claim 23.29: $\lambda \leq 2$.

To verify the claim, we first have the (obvious):

Observation 23.30: If $\mathcal{H}^r \in \text{span}\{\mathcal{H}^j : j = 1, \ldots, r-1\}$, then $(\mathcal{H}_\ell^\ell)^r \in \text{span}\{(\mathcal{H}_\ell^\ell)^j : j = 1, \ldots, r-1\}$.

Observation 23.31: For *all* k, if $(\mathcal{H}_\ell^k)^r \in \text{span}\{(\mathcal{H}_\ell^k)^j : j = 1, \ldots, r-1\}$ (where $\mathcal{H}_\ell^k$ is an appropriate submatrix of $\mathcal{H}$), then $(\mathcal{H}_\ell^{k+1})^{r+2} \in \text{span}\{(\mathcal{H}_\ell^{k+1})^j : j = 1, \ldots, r+2-1\}$. (This holds since $\mathcal{H}^{j+2}$ is a subrow of $\mathcal{H}^j$ as $h_{i,j+2} = h_{i+2,j}$.)

Let $\alpha = (\alpha_1, \alpha_2)$ be first two independent rows of $\mathcal{H}$ and note $1 \leq \alpha_1 \leq 2$. If $r \neq \alpha_1$ or α_2, then $\mathcal{H}^r \in \text{span}\{\mathcal{H}^j : j = 1, \ldots, r-1\}$ and, by (23.30), $(\mathcal{H}_\lambda^\lambda)^r \in \text{span}\{(\mathcal{H}_\lambda^\lambda)^j : j = 1, \ldots, r-1\}$. Let $\boldsymbol{\alpha}_\lambda = \{\alpha_i : i = 1, 2 \text{ and } \alpha_i \leq 2\lambda\}$ so that at least $\alpha_1 \in \boldsymbol{\alpha}_\lambda$. If $r \notin \boldsymbol{\alpha}_\lambda$ and $r \leq 2\lambda$, then $(\mathcal{H}_\lambda^\lambda)^r \in \text{span}\{(\mathcal{H}_\lambda^\lambda)^j : j = 1, \ldots, r-1\}$ and so, $2 = \rho(\mathcal{H}_\lambda^\lambda) \leq$ number of elements in $\boldsymbol{\alpha}_\lambda \leq 2$. In other words, $\boldsymbol{\alpha}_\lambda = \alpha$ and $1 \leq \alpha_1 < \alpha_2 \leq 2\lambda$. Since λ is minimal, $(\mathcal{H}_\lambda^\lambda)^{\alpha_2} \in [H_\lambda \cdots H_{2\lambda-1}]$ (otherwise $\rho(\mathcal{H}_{\lambda-1}^{\lambda-1}) = 2$) and $\rho(\mathcal{H}_{\lambda-1}^{\lambda-1}) \leq 1$. Thus, $\alpha_2 = 2(\lambda-1) + j_0$ where $j_0 = 1$ or 2. If $\lambda > 1$, then $\alpha_2 = 2(\lambda-2) + 2 + j_0$. However, in view of 23.31, if $r + 2 \in \boldsymbol{\alpha}_\lambda$, then $r \in \boldsymbol{\alpha}_\lambda$. Thus, $\alpha_1 = 2(\lambda-2) + j_0$ where $j_0 = 1$ or 2. It follows that $\lambda = 2$ (if $\lambda > 1$) since $1 \leq \alpha_1 \leq 2$. In other words, the claim is verified. *

Letting $\textbf{Hank}(2,2,2) = \{\mathcal{H} : \mathcal{H}$ a block (2×2) Hankel matrix of rank 2$\}$ and letting $\psi : \textbf{Hank}(2,2,2) \to M(4,4)$ be given by

$$\psi(\mathcal{H}) = \mathcal{H}_2^2 \qquad (23.32)$$

then we have effectively shown that $M \in \psi(\textbf{Hank}(2,2,2))$ if and only if there is an (A, B, C) in $S_{2,2}^2$ with $\textbf{H}(A, B, C) = M$. If $M_{*2}(4,4)$ is the set $\{M \in M(4,4) : \text{rank} M = 2\}$ and if G acts trivially on $M_{*2}(4,4)$, then it is easy to show that the map $(\textbf{Z}, \textbf{Y}) \to \textbf{ZY}$ is a G-isomorphism between $M_*(4,2) \times M_*(2,4)$ and $M_{*2}(4,4)$ i.e. an isomorphism between $M_*(4,2)/G \times M_*(2,4)/G$ and $M_{*2}(4,4)/G$. This generates an isomorphism between $\boldsymbol{\mathcal{P}}_{Y,Z}(S_{2,2}^2/G)$ and $\boldsymbol{\mathcal{P}}_H(S_{2,2}^2/G)$ where $\boldsymbol{\mathcal{P}}_H((A,B,C)) = \textbf{H}(A, B, C)$. In a "loose" sense, these generate a geometric quotient for the action of G. (Details and rigor are in Part II and reexamination of the "classical" proof of the ring of invariants property of a geometric quotient for scalar systems given in section 19 might prove fruitful.)

Now let us suppose that our system (A, B, C) with $m = 2, p = 2$ and $n = 2$ has an A with distinct eigenvalues λ_1, λ_2 (i.e. $\lambda_1 \neq \lambda_2$). Then $F(z) = C(zI - A)^{-1}B$ has the "partial fraction" expansion

$$F(z) = \frac{[C_1\ 0] \cdot \begin{bmatrix} B^1 \\ 0 \end{bmatrix}}{(z - \lambda_1)} + \frac{[0\ C_2] \begin{bmatrix} 0 \\ B^2 \end{bmatrix}}{(z - \lambda_2)} \qquad (23.33)$$

---

* This argument is developed in Rissanen, J. "Basis of invariants and canonical forms for linear dynamic systems", Automatica, Vol. 10, 1974.

or,

$$F(z) = \frac{C_1 B^1}{z - \lambda_1} + \frac{C_2 B^2}{z - \lambda_2} \qquad (23.34)$$

where C_1, C_2 are the columns of C and B^1, B^2 are the rows of B. In other words, (A, B, C) can be viewed as the "direct sum" of the systems $(\lambda_1, B^1, C_1), (\lambda_2, B^2, C_2)$ which have $m = 2, p = 2$ and degree $n = 1$. If, say, (A, B, C) is controllable, then so are (λ_1, B^1, C_1) and (λ_2, B^2, C_2). Conversely, if (λ_1, B^1, C_1) and (λ_2, B^2, C_2) are controllable, then so is their "direct sum". Of course, the same applies to observability. If Δ denotes the discriminant, then $(S_{2,2}^2)_\Delta$ "$=$" $S_{2,2}^1 \oplus S_{2,2}^1$. We shall use this idea shortly in proving a result concerning state feedback which is known as Heymann's lemma (and also in Part II for one proof of the Geometric Quotient Theorem).

Systems of degree $n = 1$ are also of interest in other problems. For example, in section 19, we used a "canonical form" for controllable scalar systems under state equivalence in one proof of the ring of invariants property of a geometric quotient. Such a global smooth canonical form does not exist if $m \geq 2$. Let $n = 1, m = 2$ and consider systems of the form $(a, [b_1\ b_2])$ with a, b_1, b_2 in $\mathbf{C}$. Such a system is controllable if and only if either $b_1 \neq 0$ or $b_2 \neq 0$. The action of $G = GL(1, \mathbf{C}) = \mathbf{C} - (0) = \mathbf{C}^*$ is multiplication on $[b_1\ b_2]$ and so the orbit space is simply $\mathbf{C} \times (\mathbf{C}^2 - \{0\})/\mathbf{C}^* = \mathbf{C} \times \mathbf{P}_\mathbf{C}^1$. Thus, there is no global canonical form as there is no global canonical form for the homogeneous coordinates on $\mathbf{P}_\mathbf{C}^1$. Naively, if $\xi_0 \neq 0, \xi_1 \neq 0$, then we must consider $(1, \xi_1/\xi_0)$ and $(\xi_0/\xi_1, 1)$. With more sophistication, we call a finite formal sum $\sum n_j P_j, n_j \in \mathbf{Z}$ (the integers) and $P_j \in \mathbf{P}_\mathbf{C}^1$, a *divisor* and we let $\mathrm{Div}(\mathbf{P}_\mathbf{C}^1)$ be the (free abelian) group of divisors on $\mathbf{P}_\mathbf{C}^1$. If $f = f_1/f_2$ where $f_1, f_2 \in \mathbf{C}[X_0, X_1]$ are homogeneous of the same degree and relatively prime (i.e. f is a non-zero element of the "function field", $\mathbf{C}(\mathbf{P}_\mathbf{C}^1)$ of $\mathbf{P}_\mathbf{C}^1$), then the *divisor of* f, (f) is given by

$$(f) = \sum n_i P_i \qquad (23.35)$$

where $f = \prod p_i^{n_i}, p_i$ distinct irreducible homogeneous polynomials (here p_i are linear forms) and P_i is the zero of p_i. If we define the *degree of a divisor* $D = \sum n_i P_i$ to be $\deg D = \sum n_i$, then $\deg(f) = 0$. Conversely, if $D = \sum n_i P_i$ and $\deg D = 0$, then $D = (f)$ for some f. Any such divisor is called a *principal divisor*. $\mathrm{Div}(\mathbf{P}_\mathbf{C}^1)/$ (*principal divisors*) is called the *Picard Group of* $\mathbf{P}_\mathbf{C}^1$ and denoted $\mathrm{Pic}(\mathbf{P}_\mathbf{C}^1)$.

Since the sequence $0 \to \mathbb{C}(\mathbb{P}^1_{\mathbb{C}})^*/\mathbb{C}^* \to \text{Div}(\mathbb{P}^1_{\mathbb{C}}) \overset{\text{degree}}{\longrightarrow} \mathbb{Z} \to 0$ is exact, $\text{Pic}(\mathbb{P}^1_{\mathbb{C}}) = \mathbb{Z}$. A global canonical form would in a sense correspond to a global generator of $\text{Div}(\mathbb{P}^1_{\mathbb{C}})$ modulo principal divisors so that $\text{Pic}(\mathbb{P}^1_{\mathbb{C}})$ would not be $\mathbb{Z}$. We shall see in Part II that "divisors" are important in a number of multivariable control problems.

We now turn our attention to the final result of this interlude, namely: Heymann's Lemma. Let A be $n \times n$ and let B be $n \times m$ so that the system $(A, B) \in \mathbb{A}_{\mathbb{C}}^{n^2+nm}$. If $G = GL(n, \mathbb{C})$ and if $\Gamma_f = G \times \text{Hom}_{\mathbb{C}}(\mathbb{C}^n, \mathbb{C}^m) = G \times \mathbb{A}_{\mathbb{C}}^{mn}$, then Γ_f becomes an algebraic group under the multiplication

$$[g, K][g_1, K_1] = [gg_1, Kg_1 + K_1] \tag{23.36}$$

and Γ_f acts on $\mathbb{A}_{\mathbb{C}}^{n^2+nm}$ via

$$[g, K] \cdot (A, B) = (g(A + BK)g^{-1}, gB) \tag{23.37}$$

(cf. section 21). If $\mathbf{Z}(A, B) = [B \ AB \cdots A^{n-1}B]$ is the "controllability" map, then the set of (A, B) which are not controllable is the closed set (affine) determined by the vanishing of all $n \times n$ minors of $\mathbf{Z}(A, B)$. Thus, $S^c_{n,m} = \{(A, B) : (A, B)$ is controllable $\}$ is an open set in $\mathbb{A}_{\mathbb{C}}^{n^2+nm}$. Moreover, it is invariant under the action of Γ_f. We now have:

Theorem 23.38. *(Heymann's Lemma) The pair (A, B) is controllable if and only if there is an (I, K) in Γ_f and a w in $\text{Hom}_{\mathbb{C}}(\mathbb{C}^m, \mathbb{C}) = \mathbb{C}^m$ such that the scalar system $(A + BK, Bw)$ is controllable.*

We begin with a proposition.

Proposition 23.39. *The pair (A, B) is controllable if and only if there is no proper invariant subspace $\mathbf{W}$ of A containing the columns of B i.e. $(0) < \mathbf{W} < \mathbb{A}_{\mathbb{C}}^n$, $A\mathbf{W} \subset \mathbf{W}$ and $\text{Col}(B) \subset \mathbf{W}$.*

Proof: Let $\mathbf{W}(A, B) = \text{Col}(\mathbf{Z}(A, B)) = \text{span}$ of the columns of $\mathbf{Z}(A, B)$. Then $A\mathbf{W}(A, B) \subset \mathbf{W}(A, B)$ and $\text{Col}(B) \subset \mathbf{W}(A, B)$. If rank $\mathbf{Z}(A, B) < n$, then $\mathbf{W}(A, B)$ is a subspace with the requisite properties if $B \neq 0$. If $B = 0$, a non-cyclic invariant subspace of A will do. Conversely, given such a $\mathbf{W}$, then $\text{Col}(B) \subset \mathbf{W}$ and $A\mathbf{W} \subset \mathbf{W}$

together imply that $\mathbf{W}(A, B) \subset \mathbf{W} < \mathrm{A}_{\mathbf{C}}^n$ which gives rank $\mathbf{Z}(A, B) < n$.

Corollary 23.40. *If there is a K in $\mathrm{A}_{\mathbf{C}}^{mn}$ and a w in $\mathrm{A}_{\mathbf{C}}^m$ such that $(A + BK, Bw)$ is controllable, then (A, B) is controllable.*

Proof: Note that $Bw \in \mathbf{Col}(B)$. If (A, B) were not controllable, then there is a $\mathbf{W}$ with $(0) < \mathbf{W} < \mathrm{A}_{\mathbf{C}}^n, A\mathbf{W} \subset \mathbf{W}$ and $\mathbf{Col}(B) \subset \mathbf{W}$. But $(A + BK)\mathbf{W} \subset \mathbf{W}$ for: if $v \in \mathbf{W}$, then $(A + BK)v = Av + B(Kv)$ and $Av \in \mathbf{W}$ (as $A\mathbf{W} \subset \mathbf{W}$), $B(Kv) \in \mathbf{W}$ (as $B(Kv) \in \mathbf{Col}(B)$). This contradicts the controllability of $(A + BK, Bw)$.

In other words, we have established half of Heymann's Lemma. It is the other half which is more interesting from the point of view of algebraic geometry.

Proposition 23.41. *If A has distinct eigenvalues, then (A, B) is controllable if and only if there is a $w \in \mathrm{A}_{\mathbf{C}}^m$ which is not an element of $\cup_{i=1}^n Ker(B^i)$ (i.e. (A, Bw) is controllable).*

Proof: Let $\lambda_1, \ldots, \lambda_n$ be the distinct eigenvalues of A. Then $F(z) = (zI - A)^{-1}B$ has the "partial fraction" expansion

$$F(z) = (zI - A)^{-1}B = \sum_{j=1}^n \frac{I_j \, B^j}{(z - \lambda_j)} \qquad (23.42)$$

where I_j are the columns of I_n and B^j are the rows of B. In other words, (A, B) is the "direct sum" of the systems $(\lambda_j, B^j), j = 1, \ldots, n$. It follows that (A, B) is controllable if and only if *all* (λ_j, B^j) are controllable i.e. if and only if *all* $B^j, j = 1, \ldots, n$ are non-zero. If $B^j \neq 0$ then $V_j = Ker(B^j) = \{x \in \mathrm{A}_{\mathbf{C}}^m : B^j x = 0\}$ is a proper closed subset of $\mathrm{A}_{\mathbf{C}}^m$. Hence, if *all* $B^j \neq 0$, then $\cup_{j=1}^n V_j < \mathrm{A}_{\mathbf{C}}^m$ and so, (A, B) is controllable if and only if there is a $w \notin \cup_{j=1}^n V_j$. (Note also that $(zI - A)^{-1}(Bw) = \sum_{j=1}^n I_j(B^j w)/z - \lambda_j$).

In view of Proposition 23.41, we need only prove the following proposition to establish Heymann's Lemma.

Proposition 23.43. *If (A, B) is controllable, then there is a K such that $(A + BK)$ has distinct eigenvalues.*

Proof: Following an idea of Hermann and Martin ([H-6]), we consider a map $\phi_{(A,B)} : \Gamma_f \to M(n,n)$ given by

$$\phi_{(A,B)}(g, K) = g(A + BK)g^{-1} \tag{23.44}$$

for $(g, K) \in \Gamma_f$. Let $V(\Delta)$ be the set of zeros of the discriminant Δ in $M(n,n)$ so that $M(n,n)_\Delta$ (matrices with distinct eigenvalues) is open and dense (cf. Appendix C). If the image of $\phi_{(A,B)}$, $\mathrm{Im}\,\phi_{(A,B)}$ contains an open set U, then $U \cap M(n,n)_\Delta \neq \phi$ and the proposition holds. If $\phi_{(A,B)}$ is a dominant morphism, then, in view of proposition 18.17, $\mathrm{Im}\,\phi_{(A,B)}$ contains an open set. Thus, we are reduced to proving the following lemma.

Lemma 23.45. *If (A, B) is controllable, then $\phi_{(A,B)}$ is a dominant morphism.*

Proof: We first note that Γ_f and $M(n,n) = \mathbb{A}_{\mathbb{C}}^{n^2}$ are both non-singular. Thus, letting $e = (I, 0) \in \Gamma_f$, we see that $(d\phi_{(A,B)})_e$ is a linear map of $M(n,n) \times M(m,n)(= \mathbb{A}_{\mathbb{C}}^{n^2} \times \mathbb{A}_{\mathbb{C}}^{mn})$ into $M(n,n)$ and so, $(d\phi_{(A,B)})_e$ will be surjective if and only if $(d\phi_{(A,B)})_e^*$ is injective. Note that $\langle (d\phi_{(A,B)})_e^*(M), [g, K] \rangle = \langle (d\phi_{(A,B)})_e([g, K]), M \rangle = \mathrm{tr.}\,([[(d\phi_{(A,B)})_e([g, K])]M)$. Suppose that

$$(d\phi_{(A,B)})_e([g, K]) = gA - Ag + BK \tag{23.46}$$

Then, $(d\phi_{(A,B)})_e^*(M) = 0$ would imply that

$$\mathrm{tr.}\,((gA - Ag + BK)M) = 0 \tag{23.47}$$

for all g, K. In other words,

$$\begin{aligned} \mathrm{tr.}\,((AM - MA)g) = 0 && \text{all } g \\ \mathrm{tr.}\,(MBK) = 0 && \text{all } K \end{aligned} \tag{23.48}$$

and so,

$$AM - MA = 0, \quad MB = 0 \tag{23.49}$$

or,

$$M\mathbf{Z}(A, B) = 0 \tag{23.50}$$

where $\mathbf{Z}(A, B) = [B \; AB \cdots A^{n-1}B]$. Thus, $(d\phi_{(A,B)})_e$ is surjective if and only if (A, B) is controllable (under the assumption that (23.36) holds.) But

$$
\begin{aligned}
d\phi_{(A,B)})_e(g, K) &= \frac{\partial}{\partial t}(I + tg)(A + tBK)(I + tg)^{-1}\Big|_{t=0} \\
&= \Big[(I + tg)\frac{\partial}{\partial t}\{(A + tBK)(I + tg)^{-1}\} + g(A + tBK)(I + tg)^{-1}\Big]\Big|_{t=0} \\
&= \Big[(I + tg)\{(A + tBK)(-g)(I + tg)^{-2}\} + BK(I + tg)^{-1}\Big]\Big|_{t=0} + gA \\
&= -gA + BK + gA \qquad\qquad\qquad\qquad\qquad\qquad (23.51)
\end{aligned}
$$

which is (23.46) * In fact, if $(g_0, K_0) = \gamma_0 \in \Gamma_f$, then a similar calculation shows that $(d\phi_{(A,B)})\gamma_0$ is surjective if and only if $(A + BK_0, B)$ is controllable. But $(A + BK_0, B)$ is controllable, if and only if (A, B) is controllable. So, we have shown that if (A, B) is controllable, then $(d\phi_{(A,B)})\gamma_0$ is surjective for all γ_0 in Γ_f. But $(d\phi_{(A,B)})\gamma_0$ surjective for all γ_0 in Γ_f means that the Jacobian matrix $(\partial\phi_{(A,B)}/\partial\gamma)$ has full rank n^2 globally and so $\phi^*_{(A,B)}$ must be injective. In view of proposition 7.18, $\phi_{(A,B)}$ is dominant. *

We shall prove the following general result which implies proposition 23.45 in Part II (see, e.g. [S-4]).

Proposition 23.52. *Let* $\phi : V \to W$ *be a morphism of varieties. If* $\xi \in V, \phi(\xi) \in W$ *are simple points and* $(d\phi)_\xi : T_{V,\xi} \to T_{W,\phi(\xi)}$ *is surjective, then* ϕ *is dominant.*

We hope that this brief "interlude" has provided an indication of some of the spirit of Part II: Multivariable Linear Systems and Projective Algebraic Geometry.

---

* This can be established less intuitively by algebraic means (see Part II).

* This proof of Heymann's Lemma can be found in the survey article: "Pole Assignment by Output Feedback", by C. I. Byrnes.

Appendix A. Tensor Products

Let k be a field. All vector spaces and algebras will be *over k* unless an explicit statement to the contrary is made. Let U, V, W be vector spaces. A mapping $\beta : U \times V \to W$ is *bilinear* if for each $u \in U$ the mapping $v \to \beta(u,v)$ of V into W is linear, and for each $v \in V$ the mapping $u \to \beta(u,v)$ of U into W is linear.

Definition A.1. Let U, V be vector spaces. The pair (T, β) where T is a vector space and β is a bilinear mapping of $U \times V \to T$ is a *tensor product of U and V* if : (i) span $[\beta(U \times V)] = T$; and, (ii) given any vector space W and any bilinear mapping $\beta_W : U \times V \to W$, there is a linear mapping $\ell : T \to W$ such that $\beta_W = \ell \circ \beta$. In other words, all bilinear maps factor through the tensor product (T, β).

We observe that the linear transformation ℓ of Definition A.1 is necessarily *unique* for if ℓ, ℓ' are linear mappings of T into W with $\ell \circ \beta = \ell' \circ \beta$, then $\ell - \ell' = 0$ on $\beta(U \times V)$, and, hence, by (i), on T. This leads us to the following:

Proposition A.2. *Let (T, β) and (T', β') be tensor products of U and V. Then there is a unique isomorphism $\tau : T \to T'$ such that $\tau \circ \beta = \beta'$.*

Proof: In view of the definition and our observation, there is a unique $\ell : T \to T'$ such that $\ell \circ \beta = \beta'$ and a unique $\ell' : T' \to T$ such that $\ell' \circ \beta' = \beta$. But then $\ell \circ \ell'$ and $\ell' \circ \ell$ must be the identity and so $\ell = \tau$ is the isomorphism.

In other words, the tensor product if it exists is essentially unique. We now construct a tensor product to establish existence.

Proposition A.3. *There exists a tensor product of U and V.*

Proof: Let M denote the vector space $k^{U \times V}$. The elements of M are formal finite linear combinations of elements of $U \times V$ with coefficients in k i.e. expressions of the form $\Sigma a_i(u,v)$ with $a_i \in k$. Let

N be the subspace of M generated by all elements of the forms:

$$(u_1 + u_2, v) - (u_1, v) - (u_2, v)$$
$$(u, v_1 + v_2) - (u, v_1) - (u, v_2)$$
$$(au, v) - a(u, v)$$
$$(u, av) - a(u, v)$$

where $u, u_1, u_2 \in U$, $v, v_1, v_2 \in V$ and $a \in k$. Let T be the quotient space M/N. If (u, v) is a basis element of M, then let $u \otimes v$ denote its equivalence class in T. If $\beta : U \times V \to T$ is given by $\beta(u, v) = u \otimes v$, then span $[\beta(U \times V)] = T$ and β is bilinear since $(u_1 + u_2) \otimes v = u_1 \otimes v + u_2 \otimes v$, $u \otimes (v_1 + v_2) = u \otimes v_1 + u \otimes v_2$, $(au) \otimes v = u \otimes (av) = a(u \otimes v)$. If W is any vector space and $\beta_W : U \times V \to W$ is bilinear, then β_W extends by linearity to M and vanishes on N and so generates a well-defined linear map $\ell : T \to W$ such that $\ell(u \otimes v) = \beta_W(u; v)$ i.e. $\ell \circ \beta = \beta_W$.

We shall, in view of propositions A.2 and A.3, speak of the *tensor product of U and V* which we denote by $U \otimes V$ (or $U \otimes_k V$). It is generated by the "products" $u \otimes v$. If $\{u_i\}$ is a basis of U and $\{v_j\}$ is a basis of V, then $\{u_i \otimes v_j\}$ is a basis of $U \otimes V$. In particular, if $u \neq 0$ and $v \neq 0$, then $u \otimes v \neq 0$. We note also that $U \otimes V$ is naturally isomorphic to $V \otimes U$ (via the map $u \otimes v \to v \otimes u$) and that $(U \otimes V) \otimes W$ is naturally isomorphic to $U \otimes (V \otimes W)$ (via the map $(u \otimes v) \otimes w \to u \otimes (v \otimes w)$).

Lemma A.4. *If* $u_1, \ldots, u_n$ *are linearly independent over* k *and* $\sum_{i=1}^{n} u_i \otimes v_i = 0$, *then* $v_i = 0$ *for* $i = 1, \ldots, n$.

Proof: Let $v_1', \ldots, v_r'$ be a basis of span $[v_1, \ldots, v_n]$ so that $v_i = \sum_{j=1}^{r} a_{ij} v_j'$. Then $\sum_{i=1}^{n} \sum_{j=1}^{r} a_{ij}(u_k \otimes v_j') = 0$ and all the $a_{ij} = 0$ so that $v_i = 0$ for $i = 1, \ldots, n$.

If $\phi : U \to U_1$ and $\psi : V \to V_1$ are linear maps, then the map $\phi \otimes \psi : U \otimes V \to U_1 \otimes V_1$ given by

$$(\phi \otimes \psi)(u \otimes v) = \phi(u) \otimes \psi(v) \qquad (A.5)$$

is linear. If ϕ and ψ are surjective, then so is $\phi \otimes \psi$. If ϕ and ψ are injective, then $\phi \otimes \psi$ is also injective (for if $\{u_i\}$ is a basis of U and $\{v_j\}$ is a basis of V, then $(\phi \otimes \psi)(u_i \otimes v_j) = \phi(u_i) \otimes \psi(v_j)$ is linearly independent in $U_1 \otimes V_1$ as $\Sigma a_{ij}\phi(u_i) \otimes \psi(v_j) = \Sigma_i \phi(u_i) \otimes (\psi(\Sigma_j a_{ij}v_j))$ and so $(\phi \otimes \psi)(\Sigma a_{ij}u_i \otimes v_j) = 0$ implies $\Sigma a_{ij}\phi(u_i) \otimes \psi(v_j) = 0$ and hence that $a_{ij} = 0$, for all i, j.)

Lemma A.6. *Let U_1 be a subspace of U and let V_1 be a subspace of V. Let $\phi_1 : U \to U/U_1$, $\psi_1 : V \to V/V_1$ be the natural linear maps. Then $(U/U_1) \otimes (V/V_1)$ is isomorphic to $(U \otimes V)/[U_1, V_1]$ where $[U_1, V_1] = U_1 \otimes V + U \otimes V_1$ and hence, $\mathrm{Ker}(\phi_1 \otimes \psi_1) = [U_1, V_1]$.*

Proof: We shall show that $(U \otimes V)/[U_1, V_1]$ is in fact a tensor product of U/U_1 and V/V_1. We write $\phi_1(u) = \overline{u}$ and $\psi_1(v) = \overline{v}$ and we let $\omega : U \otimes V \to U \otimes V/[U_1, V_1]$ be the natural linear map. We observe that the map $(u, v) \to \omega(u \otimes v)$ is bilinear on $U \times V$ and is 0 for either u in U_1 or v in V_1 (by the definitions of ω and $[U_1, V_1]$). Therefore, $\omega(u \otimes v)$ depends only on the classes $\overline{u}$ and $\overline{v}$ i.e. $\omega(u \otimes v)$ may be written as $\overline{\omega}(\overline{u}, \overline{v})$. Setting $W = U \otimes V/[U_1, V_1]$, the map $\overline{\omega} : (U/U_1) \times (V/V_1) \to W$ given by $\overline{\omega}(\overline{u}, \overline{v}) = \omega(u \otimes v)$ is bilinear and clearly satisfies $W = \mathrm{span}\,[\overline{\omega}(U/U_1 \times V/V_1)]$. Let $\tilde{W}$ be any vector space (over k) and let $\tilde{\beta} : U/U_1 \times V/V_1 \to \tilde{W}$ be a bilinear map. Then for any $u \in U$, $v \in V$, the map $\beta : U \times V \to \tilde{W}$ given by $\beta(u, v) = \tilde{\beta}(\overline{u}, \overline{v})$ is bilinear and there exists a linear map $\ell : U \otimes V \to \tilde{W}$ with $\ell(u \otimes v) = \beta(u, v) = \tilde{\beta}(\overline{u}, \overline{v})$. But then ℓ is 0 for either u in U_1 or v in V_1 and so ℓ is 0 on $[U_1, V_1]$. It follows that $\tilde{\ell} = \ell \circ \omega$ with $\ell : W \to \tilde{W}$ a linear map. Then $\tilde{\beta}(\overline{u}, \overline{v}) = \ell(\omega(u \otimes v)) = \ell(\overline{\omega}(\overline{u}, \overline{v}))$ or in other words, $\tilde{\beta} = \ell \circ \overline{\omega}$. Thus, $U \otimes V/[U_1, V_1]$ is a tensor product of (U/U_1) and (V/V_1) and the result follows by proposition A.2.

Now let R and S be k-algebras i.e. k is a subring of R and of S and the element 1 of k is also the unit 1_R of R and the unit 1_S of S. Then $R \otimes_k S$ is defined as a k-vector space. However, we can introduce a multiplication on the tensor product $R \otimes_k S$: for elements $(r \otimes s)$ and $(r_1 \otimes s_1)$ it is given by

$$(r \otimes s)(r_1 \otimes s_1) = (rr_1) \otimes (ss_1) \tag{A.7}$$

and in general by

$$(\Sigma_i(r_i \otimes s_i))(\Sigma_j(r'_j \otimes s'_j)) = \Sigma_{i,j}(r_i r'_j \otimes s_i s'_j) \tag{A.8}$$

It is easy to see that $R \otimes_k S$ becomes a commutative ring with identity $1 \otimes 1$. We also observe that the mappings $\psi_R : R \to R \otimes S$ and

$\psi_S : S \to R \otimes S$ given by $\psi_R(r) = r \otimes 1$ and $\psi_S(s) = 1 \otimes s$ are injective k-homomorphisms so that R and S may be viewed as subrings of $R \otimes_k S$. We shall adopt this point of view here. Finally we note that the mapping $a \to a(1 \otimes 1) = a \otimes 1 = 1 \otimes a$ is a natural imbedding of k into $R \otimes_k S$. Thus, $R \otimes_k S$ is in fact a k-algebra which we call the *tensor product of the k-algebras R and S*.

Example A.9.   If $k \subset K$ with K a field, then $K \otimes_k S$ contains the field K and any basis of S over k is also a basis of $K \otimes_k S$ over K. Thus $k \otimes_k S = S$. Alternatively, the map $a \otimes s \to as$ is an isomorphism between $k \otimes_k S$ and S.

Example A.10.   Let $R = k[X_1, \ldots, X_N]$ and $S = k[Y_1, \ldots, Y_M]$ be polynomial rings. Then $R \otimes_k S \simeq k[X_1, \ldots, X_N, Y_1, \ldots, Y_M]$, for the mapping $X_i \otimes 1 \to X_i, 1 \otimes Y_j \to Y_j$ clearly generates a k-isomorphism.

Proposition A.11.   *Let R and S be k-algebras and let $\mathfrak{a}$ be an ideal in R and $\mathfrak{b}$ be an ideal in S. If $[\mathfrak{a}, \mathfrak{b}] = \mathfrak{a} \otimes S + R \otimes \mathfrak{b}$, then $(R/\mathfrak{a}) \otimes (S/\mathfrak{b})$ is isomorphic to $R \otimes S/[\mathfrak{a}, \mathfrak{b}]$.*

Proof: Let $\phi_1 : R \to R/\mathfrak{a}$, $\psi_1 : S \to S/\mathfrak{b}$ be the natural maps. Since ϕ_1, ψ_1 are surjective k-homomorphisms, $\phi_1 \otimes \psi_1 : R \otimes S \to (R/\mathfrak{a}) \otimes (S/\mathfrak{b})$ is a surjective k-homomorphism. However, lemma A.6 implies that $\text{Ker}(\phi_1 \otimes \psi_1) = [\mathfrak{a}, \mathfrak{b}]$ and the proposition follows.

We note that if $\{r_i\}$ is a set of generators of R and $\{s_j\}$ is a set of generators of S, then the set $\{r_i \otimes 1, 1 \otimes s_j\}$ is a set of generators for $R \otimes S$. In particular, if R and S are finitely generated, then so is their tensor product $R \otimes_k S$.

Proposition A.12.   *If R and S are affine k-algebras, then $R \otimes_k S$ is an affine k-algebra.*

Proof: Since $R \otimes_k S$ is finitely generated, it will be enough to show that there are no non-zero nilpotents. Let $\sum_{i=1}^{n} r_i \otimes s_i$ be nilpotent and assume without loss of generality that the s_i are linearly independent over k. Let $\alpha \in \text{Hom}(R, k)$ and consider the map $\alpha \otimes 1 : R \otimes S \to S$ given by $(\alpha \otimes 1)(r \otimes s) = \alpha(r)s$. Since $\alpha \otimes 1 \in \text{Hom}_k(R \otimes S, S)$, $\sum_{i=1}^{n} \alpha(r_i)s_i$ is a nilpotent in S. But S is an affine k-algebra and so $\sum_{i=1}^{n} \alpha(r_i)s_i = 0$. Since the s_i are linearly independent over k,

$\alpha(r_i) = 0$ for $i = 1, \ldots, n$. It follows from the Nullstellensatz (cf. problem 20) that $r_i = 0$ for $i = 1, \ldots, n$.

Proposition A.13. *Let R be an affine k-algebra. Suppose that R and S are integral domains. Then $R \otimes_k S$ is an integral domain.*

Proof: Let $u = \Sigma_i r_i \otimes s_i$ and $u' = \Sigma_j r'_j \otimes s'_j$ be elements of $R \otimes S$ such that $uu' = 0$ and assume without loss of generality that the s_i and s'_j are linearly independent over k. Let $\alpha \in \text{Hom}(R, k)$ and again consider the map $\alpha \otimes 1 : R \otimes S \to S$ given by $(\alpha \otimes 1)(r \otimes s) = \alpha(r)s$. Since $\alpha \otimes 1 \in \text{Hom}_k(R \otimes S, S)$, we have

$$(\alpha \otimes 1)(uu') = (\alpha \otimes 1)(u) \cdot (\alpha \otimes 1)(u') = (\Sigma_i \alpha(r_i)s_i)(\Sigma_j \alpha(r'_j)s'_j) = 0.$$

But S is an integral domain and the s_i and s'_j are linearly independent over k and so, either $\alpha(r_i) = 0$ for all i or $\alpha(r'_j) = 0$ for all j. It follows that $\alpha(r_i r'_j) = 0$ for all i, j and hence, as α was arbitrary, that $r_i r'_j = 0$ for *all* i, j. If, say, $u \neq 0$, then some r_i is not zero and $r'_j = 0$ for all j. In other words, if $u \neq 0$, then $u' = 0$. Thus $R \otimes_k S$ is an integral domain.

Corollary A.14. *Let R and S be k-algebras and let $\mathfrak{a}$ be a prime ideal in R and $\mathfrak{b}$ be a prime ideal in S. Then $[\mathfrak{a}, \mathfrak{b}] = \mathfrak{a} \otimes S + R \otimes \mathfrak{b}$ is a prime ideal in $R \otimes S$.*

Corollary A.15. *Let R and S be integral domains and let $f \neq 0$ be an element of R. Then $R_f \otimes_k S \cong (R \otimes_k S)_{f \otimes 1}$.*

Proof: Consider the map $\psi : R_f \otimes S \to (R \otimes S)_{(f \otimes 1)}$ given by $\psi(r/f \otimes s) = (r \otimes s)/(f \otimes 1)$ and extended in the natural way. ψ is clearly a k-homomorphism. If $h = \Sigma_i (r_i \otimes s_i)/(f \otimes 1)$, then $\psi(\Sigma_i r_i/f \otimes s_i) = h$ so that ψ is surjective. Since R and S are integral domains, so are $R \otimes S$ and $(R \otimes S)_{(f \otimes 1)}$. If $\psi(\Sigma_i r_i/f \otimes s_i) = 0$ with the s_i linearly independent over k, then $\Sigma_i (r_i \otimes s_i)/(f \otimes 1) = 0$ and so $\Sigma_i (r_i \otimes s_i) = 0$. The injectivity of ψ follows from lemma A.4.

Corollary A.16. *Let R and S be integral domains and let $f \neq 0$, $g \neq 0$ be elements of R, S respectively. Then $(R_f \otimes_k S_g) = (R \otimes_k S)_{f \otimes g}$.*

Proof: Note that $R_f \otimes S_g = (R \otimes S_g)_{(f \otimes 1)} = [(R \otimes S)_{(1 \otimes g)}]_{(f \otimes 1)} = (R \otimes S)_{(f \otimes 1)(1 \otimes g)} = (R \otimes S)_{(f \otimes g)}$.

We shall now extend the notion of a tensor product to "modules".

Definition A.17. Let R be a ring. An abelian group M on which R acts linearly is called an *R-module* (i.e. M is an R "vector" space). More explicitly, there is a mapping $R \times M \to M$, $(r, m) \to rm$, such that $r(m_1 + m_2) = rm_1 + rm_2$, $(r + r_1)m = rm + r_1 m$, $(rr_1)m = r(r_1 m)$, and $1m = m$.

If L, M, N are R-modules, then a mapping $\beta : M \times N \to L$ is *R-bilinear*, if for each $m \in M$, the mapping $n \to \beta(m, n)$ of N into L is R-linear, and for each $n \in N$, the mapping $m \to \beta(m, n)$ of M into L is R-linear. The tensor product of two R-modules M and N can be developed in a manner similar to that used for vector spaces using the universal property for factoring bilinear maps. More precisely, we have:

Theorem A.18. *Let M, N be R-modules. Then there is an R-module T and a bilinear mapping $\beta_T : M \times N \to T$ such that : (*) if L is any R-module and β is any bilinear mapping of $M \times N$ into L, then there is a unique R-linear mapping $\ell_\beta : T \to L$ such that $\beta = \ell_\beta \circ \beta_T$. Moreover, if T' and $\beta_{T'}$ satisfy (*), then there is a unique isomorphism $\tau : T \to T'$ such that $\tau \circ \beta_T = \beta_{T'}$.*

Proof: Uniqueness is proved in a manner entirely similar to that used in proposition A.2. As for existence, we consider the free R-module $R^{M \times N}$ i.e. the expressions $\sum_{i=1}^n r_i(m_i, n_i)$. If P is the sub-module of $R^{M \times N}$ generated by all elements of the following forms:

$$(m + m_1, n) - (m, n) - (m_1, n)$$
$$(m, n + n_1) - (m, n) - (m, n_1)$$
$$(rm, n) - r(m, n)$$
$$(m, rn) - r(m, n)$$

then we let T be the quotient R-module $R^{M \times N}/P$ and for each basis element (m, n) in $R^{M \times N}$, we let $m \otimes n$ be its residue class in T. Then T is generated by the elements $m \otimes n$ and the mapping $\beta_T : M \times N \to T$

defined by $\beta_T(m, n) = m \otimes n$ and extended by linearity, is clearly R-bilinear. If L is an R-module and $\beta : M \times N \to L$ is bilinear, then the map $\ell_\beta : T \to L$ given by $\ell_\beta(m \otimes n) = \beta(m, n)$ satisfies $(*)$.

Definition A.19.   The R-module T of theorem A.18 is called the *tensor product of M and N* and is denoted by $M \otimes_R N$ or simply by $M \otimes N$.

We note that R is itself an R-module and that $R \otimes_R N$ " $=$ " N in the sense that the mapping $r \otimes n \to rn$ is an isomorphism. We also have

Proposition A.20. *If M and N are R-modules, if M' is a submodule of M and if N' is a submodule of N, then*

$$(M/M') \otimes_R (N/N') \simeq M \otimes_R N / [\mathrm{Im}(M' \otimes_R N) + \mathrm{Im}(M \otimes_R N')] \quad (A.21)$$

under a canonical isomorphism.

Proof: (see [F-2]).

Appendix B ([M-4]). Actions of Reductive Groups

Let G be an affine algebraic group over k. (This means that G is an affine algebraic set which is a group and that the group operations are k-morphisms.) Let $S = k[G]$ be the ring of regular functions on G. We note that S is a k vector space. Let M be a k vector space and let σ be a linear map of M into $k[G] \otimes M$ so that

$$\sigma(m) = \Sigma f_i \otimes m_i \qquad (\text{B.1})$$

with $f_i \in k[G]$. Then σ defines a mapping $\alpha_\sigma : G \times M \to M$ by setting

$$\alpha_\sigma(g, m) = \Sigma f_i(g) m_i \qquad (\text{B.2})$$

where $\sigma(m)$ is given by (B.1). We say that σ defines an *action of G on M* if α_σ is an action (i.e. $\alpha_\sigma(g_1 g_2, m) = \alpha_\sigma(g_1, \alpha_\sigma(g_2, m))$ and $\alpha_\sigma(\epsilon, m) = m$), and, in that case, we say that *G acts on M*. If σ defines an action, then $m \in M$ implies m is contained in a finite dimensional invariant subspace ($\Sigma k m_i$). Clearly, m is invariant if and only if $\sigma(m) = 1(\cdot)m$ and $M_1 \subset M$ is invariant if and only if $\sigma(M_1) \subset k[G] \otimes M_1$. We usually suppress the linear map σ.

Example B.3. Let M be n-dimensional with $e_1, \ldots, e_n$ as basis. If $\sigma(e_i) = \Sigma_j f_{ij}(\cdot) e_j$, then $(f_{ij}(\cdot)) \in GL(n, k)$ and the map $g \to (f_{ij}(g))$ is a k-homomorphism of G into $GL(n, k)$.

Definition B.4. Let G act on a finite dimensional vector space M. The action is *completely reduced* if, for any invariant subspace M_1 of M, there is an invariant complement (i.e. an invariant subspace M_2 of M such that $M = M_1 \oplus M_2$). The group G is *reductive* if each action on any M is completely reduced.

The group $GL(n, k)$, k of characteristic 0, is an important example of a reductive group ([D-2]). The importance of reductivity for our purposes lies in the following results.

Proposition B.5. *If G is reductive and acts on M, then the subspace $M^G = \{m : g \cdot m = m\}$ of invariant elements has a unique invariant complement M_G.*

Proof: By Zorn's lemma, there is a maximal invariant subspace M_G such that $M_G^G = M_G \cap M^G = \{0\}$. If N is any invariant subspace of M and $m \in N$, then there is a finite dimensional invariant subspace $N_1 \subset N$ with $m \in N_1$, and hence, an invariant subspace $N' \subset N_1$ with $N_1 = (N_1 \cap M_G) \oplus N'$ (as G is reductive and N_1 is finite dimensional). If $N^G = \{0\}$, then $N'^G = \{0\}$ and $(M_G \oplus N')^G = (M_G \oplus N') \cap M^G = \{0\}$. By the maximality of M_G, we have $N' = \{0\}$ and so, $N \subset M_G$ if $N^G = \{0\}$. In other words, M_G is unique. Since $M_G \cap M^G = \{0\}$, we need only show that $M = M^G \oplus M_G$. So let $m \in M$ and let M_1 be a finite dimensional invariant subspace with $m \in M_1$. Since G is reductive, there is an invariant $M_1' \subset M_1$ with $M_1 = (M_1 \cap M^G) \oplus M_1'$ and $M_1'^G = \{0\}$. This implies that $M_1' \subset M_G$ and $m \in M_1 \subset M^G \oplus M_G$.

Definition B.6. If G is reductive and acts on M, then the mapping $P_M : M \to M^G$ is called the *Reynolds operator on M*.

We note that P_M is a projection of M onto M^G with kernel M_G.

Proposition B.7. *If G is reductive and acts on M and N and if $L : M \to N$ is a G-linear map (i.e. commutes with the action of G), then*

$$P_N L = L P_M \qquad\qquad (B.8)$$

i.e. L commutes with the Reynolds operator.

Proof: First note that $L(M^G) \subset N^G$ since, if $g \cdot m = m$, then $L(m) = L(g \cdot m) = g \cdot L(m)$. So it is enough to show that $L(M_G) \subset N_G$. If $m \in M_G$, then there is an $M_1 \subset M_G$ with $m \in M_1$ and $M_1 = (M_1 \cap \mathrm{Ker}\,L) \oplus M_1'$ where M_1' is invariant. Since $L(M_1) = L(M_1')$, $L(M_1)^G = L(M_1')^G = L(M_1'^G) = \{0\}$ (as $M_1'^G \subset M_G \cap M^G$). It follows that $L(m) \in L(M_1') \subset N_G$.

Corollary B.9. *Let R be a k-algebra and suppose that the reductive group G acts on R by k-automorphisms (i.e. the map $r \to g \cdot r$, $g \in G$ is a k-automorphism of R). Then*

$$P_R(xy) = x P_R(y) \qquad \text{(Reynolds Identity)} \qquad (B.10)$$

if $x \in R^G$ and $y \in R$.

Proof: The mapping $y \to xy$ is G-linear if x is an element of R^G.

Corollary B.11. *Let R be a k-algebra and suppose that the reductive group G acts on R by k-automorphisms. If $\mathfrak{a}_1, \ldots, \mathfrak{a}_r$ are invariant ideals in R, then $(\sum_{i=1}^r \mathfrak{a}_i) \cap R^G = \sum_{i=1}^r (\mathfrak{a}_i \cap R^G)$.*

Proof: Since $\Sigma(\mathfrak{a}_i \cap R^G) \subset (\Sigma \mathfrak{a}_i) \cap R^G$, we need only show that if $f = \Sigma f_i$ with $f_i \in \mathfrak{a}_i$ and $f \in R^G$, then $f = \Sigma f_i'$ with $f_i' \in \mathfrak{a}_i \cap R_G$. Let P_R be the Reynolds operator on R and let $P_i = P_R$ restricted to $\mathfrak{a}_i$ be the Reynolds operator on $\mathfrak{a}_i$ (noting that $\mathfrak{a}_i$ is invariant). Then

$$f = P_R f = \Sigma P_R f_i = \Sigma P_i f_i \qquad (B.12)$$

since $f \in R^G$. But $P_i f_i = f_i' \in \mathfrak{a}_i^G \subset \mathfrak{a}_i \cap R^G$ and the corollary is established.

Let us suppose that R is a k-algebra and that the reductive group G acts on R by k-automorphisms. If S is an R^G-algebra, then G also acts on $R \otimes_{R^G} S$ via

$$g \cdot (r \otimes s) = (g \cdot r) \otimes s \qquad (B.13)$$

where $g \in G$, $r \in R$ and $s \in S$. We then have:

Lemma B.14. *If S is an R^G-algebra, then*

$$(R \otimes_{R^G} S)^G = S \qquad (B.15)$$

i.e. the ring of invariants in $(R \otimes_{R^G} S)$ is S.

Proof: Let $P = P_R$ and $P' = P_{R \otimes_{R^G} S}$ be the Reynolds operators. Since $S \simeq R^G \otimes_{R^G} S$ (appendix A) $\subset (R \otimes_{R^G} S)^G$, it will be enough to show that if $\Sigma(r_i \otimes s_i) \in (R \otimes_{R^G} S)^G$, then $\Sigma(r_i \otimes s_i) \in R^G \otimes_{R^G} S$. But $P'(\Sigma(r_i \otimes s_i)) = P'(\Sigma(r_i \otimes 1)(1 \otimes s_i)) = \Sigma[P'(r_i \otimes 1)](1 \otimes S_i)$ (by Reynolds identity). Since $P'(r_i \otimes 1) = (Pr_i) \otimes 1$ by proposition B.7, we have $\Sigma(r_i \otimes s_i) = P'(\Sigma(r_i \otimes s_i)) = \Sigma(Pr_i) \otimes s_i \in R^G \otimes_{R^G} S$ and the lemma holds.

Corollary B.16. *If $\mathfrak{a}$ is an ideal in R^G, then $R\mathfrak{a} \cap R^G = \mathfrak{a}$.*

Proof: We observe that $R \otimes_{R^G} R^G/\mathfrak{a} = R/R\mathfrak{a}$ in view of proposition A.20. It follows from the lemma that $(R/R\mathfrak{a})^G = R^G/\mathfrak{a}$ and hence that $R\mathfrak{a} \cap R^G = \mathfrak{a}$.

Corollary B.17. *If R is noetherian, then R^G is noetherian.*

Proof: By corollary B.16, the mapping $\mathfrak{a} \to \mathfrak{a}^e = R\mathfrak{a}$ is an injection and satisfies the condition that $\mathfrak{a} < \mathfrak{a}_1$ implies $R\mathfrak{a} < R\mathfrak{a}_1$. Thus, if $\{\mathfrak{a}_i\}$ is an ascending chain of ideals in R^G, then $\{R\mathfrak{a}_i\}$ is an ascending chain of ideals in R. Since R is noetherian, $R\mathfrak{a}_\nu = R\mathfrak{a}_{\nu+1} = \cdots$ for some ν and so, $\mathfrak{a}_\nu = \mathfrak{a}_{\nu+1} = \cdots$.

A ring R is a *graded ring* if there is a family $R_n, n = 0, 1, \ldots$ of subgroups of the additive group of R such that (i) $R = \oplus\Sigma R_n$ and (ii) $R_m R_n \subseteq R_{m+n}$ for all $m, n \geq 0$. In that case, R_0 is a subring of R and each R_n is an R_0 module. Any element r of R can be written uniquely as a sum Σr_n where $r_n \in R_n$ and all but a finite number of r_n are 0. The non-zero r_n are called the *homogeneous* components of r. Elements of R_n are called *homogeneous of degree n*.

Example B.18. Let $R = k[X_1, \ldots, X_N]$ and let R_n be the set of all homogeneous polynomials of degree n.

Proposition B.19. *If R is a graded noetherian ring, then R_0 is noetherian and R is a finitely generated R_0-algebra (i.e. $R = R_0[f_1, \ldots, f_N]$).*

Proof: Let $R_+ = \oplus\sum_{n>0} R_n$. Then R_+ is an ideal in R and $R_0 \simeq R/R_+$. It follows that R_0 is noetherian and that R_+ has a finite basis $f_1, \ldots, f_N$ which we may assume are homogeneous of degrees $d_1, \ldots, d_N$ (all > 0). Let $R' = R_0[f_1, \ldots, f_N] \subset R$. We claim that $R' = R$. To verify the claim, we need only show that $R_n \subset R'$ for all n. We use induction on n. Clearly $R_0 \subset R'$. Let $n > 0$ and let $r \in R_n$. Since $r \in R_+$, $r = \sum_{i=1}^N r_i f_i$ where $r_i \in R_{n-d_i}$ or $r_i = 0$ (if $n < d_i$). By induction, $r_i \in R'$ and so $r \in R'$.

Let us suppose now that R is a graded noetherian k-algebra (with $R_0 = k$) and that the reductive group G acts homogeneously (i.e. $G \cdot R_n \subset R_n$) on R by k-automorphisms. Then $R^G = \oplus\sum_{n=0}^\infty R_n^G$ is also a graded noetherian k-algebra (with $R_0^G = k$) and is, by proposition B.19, a finitely generated k-algebra. In other words, in this situation, the *equivalence E_G on R is proper* (definition 14.5). Thus, if k is of characteristic 0, then the equivalence E_G, $G = GL(n, k)$, on R is proper. If k has characteristic p, then not every action on any M is completely reduced for $GL(n, k)$. Nonetheless, ([M-3], [H-1], [S-4]), the equivalence E_G is still proper.

Appendix C. Symmetric Functions and Actions of the Symmetric Group

Let us consider the set of integers $\{1, \ldots, n\}$. A bijective map of this set onto itself is called a *permutation* on n letters. The set of all such permutations forms a group S_n called the *symmetric group of degree n*. The order of S_n (i.e. the number of elements) is $n!$. An element σ of S_n such that

$$\sigma(i_1) = i_2 \quad , \quad \sigma(i_2) = i_3, \ldots, \sigma(i_r) = i_1 \tag{C.1}$$

and leaves the rest of $\{1, \ldots, n\}$ fixed is called a *cycle* and we write $\sigma = (i_1 i_2 \cdots i_r)$. Every $\sigma \in S_n$ can be written as a product of disjoint cycles. A cycle of the form $(i_1 i_2)$ is called a *transposition*. Since $(i_1 \cdots i_r) = (i_1 i_2)(i_1 i_3) \cdots (i_1 i_r)$, every $\sigma \in S_n$ is a product of transpositions.

Let $R = k[x_1, \ldots, x_n]$. Then we may define an action of S_n on R via automorphisms as follows:

$$(\sigma \cdot f)(x_1, \ldots, x_n) = f^\sigma(x_1, \ldots, x_n) = f(x_{\sigma(1)}, \ldots, x_{\sigma(n)}) \tag{C.2}$$

where $f \in R$ and $\sigma \in S_n$.

Definition C.3. A polynomial function f in R is *symmetric* if $\sigma \cdot f = f$ for all σ in S_n i.e. f is an invariant for the action of S_n on R. We let R^{S_n} be the *ring of symmetric functions*. The polynomials

$$\psi_r(x_1, \ldots, x_n) = \sum_{1 \le i_1 < \cdots < i_r \le n} x_{i_1} x_{i_2} \cdots x_{i_r} \tag{C.4}$$

$r = 1, \ldots, n$ are called *elementary symmetric polynomials*.

Proposition C.5. *The ψ_r are symmetric.*

Proof: Now let z be an indeterminate over R and let $F(z) = (z - x_1)(z - x_2) \cdots (z - x_n)$. The action of S_n on R can be extended to an action on $R[z]$ by setting $\sigma(z) = z$. Then $\sigma \in S_n$ permutes the factors of F so that $\sigma \cdot F = F$. Thus, σ leaves the coefficients of F invariant. But $F(z) = z^n - \psi_1(x)z^{n-1} + \psi_2(x)z^{n-2} - \cdots + (-1)^n \psi_n(x)$ so that the $\psi_r(x)$ are symmetric.

Proposition C.6. *R is algebraic over $k[\psi_1, \ldots, \psi_n]$.*

Proof: Simply note that $F(x_i) = x_i^n - \psi_1(x)x_i^{n-1} + \cdots + \psi_n(x) = 0$.

Proposition C.7. *$R^{S_n} = k[\psi_1, \ldots, \psi_n]$.*

Proof: If $f \in R^{S_n}$, then $f = f_0 + f_1 + \cdots + f_m$ where f_i is homogeneous of degree i and, since the action of S_n preserves degrees, each f_i is in R^{S_n}. Thus, we may suppose f is homogeneous of degree m. Call a monomial $x_1^{k_1} \cdots x_n^{k_n}$ of degree m greater than $x_1^{\ell_1} \cdots x_n^{\ell_n}$ if $k_1 = \ell_1, \ldots, k_s = \ell_s$ but $k_{s+1} > \ell_{s+1}$. For example, $x_1^2 x_2 x_3 > x_1 x_2^3 > x_1 x_2^2 x_3$. Let $a x_1^{k_1} \cdots x_n^{k_n}$ be the greatest term in f. Since f is symmetric, $k_1 \geq k_2 \geq \cdots \geq k_n$. The greatest term in $a\psi_1^{k_1-k_2}\psi_2^{k_2-k_3} \cdots \psi_n^{k_n}$ is thus the same as that of f. But then $f_1 = f - a\psi_1^{k_1-k_2} \cdots \psi_n^{k_n}$ is a homogeneous symmetric polynomial whose greatest term is less than that of f. By induction, $f_1 \in k[\psi_1, \ldots, \psi_n]$.

Proposition C.8. *The ψ_r are algebraically independent over k.*

Proof 1. If there is a non-trivial relation over k amongst the ψ_r, then distinct terms in the ψ_r have distinct greatest terms in the x_j. The greatest term in the x_j occurs only once which gives a non-trivial relation amongst the x_j (a contradiction).

Proof 2. We observe that

$$\psi_1(x_1, \ldots, x_n) = x_1 + \psi_1'(x_2, \ldots, x_n)$$
$$\psi_2(x_1, \ldots, x_n) = x_1\psi_1'(x_2, \ldots, x_n) + \psi_2'(x_2, \ldots, x_n)$$
$$\vdots$$
$$\psi_n(x_1, \ldots, x_n) = x_1\psi_{n-1}'(x_2, \ldots, x_n)$$

(C.9)

where the ψ_j' are the elementary symmetric polynomial in $x_2, \ldots, x_n$. If $P(\psi_1, \ldots, \psi_n) = 0$ and we set $x_1 = 0$, then $P(\psi_1', \ldots, \psi_{n-1}', 0) = 0$. By induction, the ψ_j' are algebraically independent. Thus, $P(z_1, \ldots, z_n) = z_n P_1(z_1, \ldots, z_n)$ with P_1 of lower degree than P. Since $\psi_n \neq 0$, $P_1(\psi_1, \ldots, \psi_n) = 0$. By induction on the degree, $P_1 \equiv 0$.

Proposition C.10. *Let* $s_r(x_1, \ldots, x_n) = \sum_{i=1}^{r} x_i^r$, $r = 1, \ldots, n$. *Then the* s_r *are symmetric and, for* k *of characteristic 0, satisfy Newton's identities*

$$s_r - \psi_1 s_{r-1} + \psi_2 s_{r-2} \cdots + (-1)^r r \psi_r = 0 \qquad (C.11)$$

for $r = 1, \ldots, n$.

Proof: Clearly the s_r are symmetric. As for the rest, simply use induction and (C.9).

Example C.12. Let $r = 3$ and $s_i' = \sum_{j=2}^{n} x_j^i$. Then $s_3 - \psi_1 s_2 + \psi_2 s_1 - \psi_3 = x_1^3 + s_3' - (x_1 + \psi_1')(x_1^2 + s_2') + (x_1 \psi_1' + \psi_2')(x_1 + s_1') - x_1 \psi_2' - \psi_3' = s_3' - \psi_1' s_2' + \psi_2' s_1' - \psi_3' - x_1(s_2' - \psi_1' s_1') = 0$ (by induction on n).

If $A \in M(n, n, k)$ is an $n \times n$-matrix, then

$$\det(zI - A) = z^n - \chi_1(A)z^{n-1} - \cdots - \chi_n(A) \qquad (C.13)$$

where the $\chi_i(A)$ are the *characteristic coefficients of* A. The roots of $\det(zI - A) = 0$ are the *eigenvalues of* A. If $\lambda_1, \ldots, \lambda_n$ are the eigenvalues of A, then $\chi_i(A) = (-1)^{i-1} \psi_i(\lambda_1, \ldots, \lambda_n)$. The characteristic coefficients can be computed directly from A and, in particular, we have, for characteristic $k = 0$:

Proposition C.14. *Let* $s_r(A) = \operatorname{tr}(A^r)$ *(the trace of* A^r*) for* $r = 1, \ldots, n$. *Then*

$$s_r(A) - \chi_1(A)s_{r-1}(A) - \cdots - r\chi_r(A) = 0 \qquad (C.15)$$

for $r = 1, \ldots, n$.

Proof: If $\lambda_1, \ldots, \lambda_n$ are the eigenvalues of A, then $\operatorname{tr}(A^r) = \Sigma \lambda_i^r$ and apply proposition C.10.

Definition C.16. Let $D(x_1, \ldots, x_n) = \prod_{1 \leq i < j \leq n}(x_i - x_j)^2$. Then D is symmetric and so $D(x_1, \ldots, x_n) = \Delta(\psi_1, -\psi_2, \ldots, (-1)^{n-1}\psi_n)$. The function $\Delta(A) = \Delta(\chi_1(A), \ldots, \chi_n(A))$ is called the *discriminant of* A.

Proposition C.17. $A \in M(n, n, k)$ *has distinct eigenvalues if and only if* $\Delta(A) \neq 0$.

Proof: A has distinct eigenvalues if and only if $D(\lambda_1, \ldots, \lambda_n) \neq 0$ which holds if and only if $\Delta(A) \neq 0$.

Corollary C.18. *The set* $\{A \in M(n, n, k) : A$ *has distinct eigenvalues*$\}$ *is open and dense.*

Proof: Simply observe that the set is the principal affine open set $(\mathbb{A}_k^{n^2})_\Delta$.

Proposition C.19.   *If* $g \in GL(n, k)$ *and* $A_1 = gAg^{-1}$, *then* $\Delta(A_1) = \Delta(A)$.

Proof: $\det(zI - A_1) = \det(zgg^{-1} - gAg^{-1})$
$$= \det g \det(zI - A) \det g^{-1} = \det(zI - A).$$
Now let $D(n, k)$ be the set of $n \times n$ diagonal matrices with entries in k and let

$$N = \{g \in GL(n, k): \quad gD(n, k)g^{-1} = D(n, k)\} \tag{C.20}$$

Clearly, N is a subgroup of $GL(n, k)$ and is, in fact, the set of so-called *monomial matrices* (i.e. matrices with exactly one non-zero entry in each row and column). We also note that N contains a subgroup which is isomorphic to the symmetric group S_n. To see this, we observe that it is enough to consider the transpositions (ij) and that, if $j > i$, then $(ij) = (ii + 1)(i + 1 \ i + 2) \cdots (j - 1 \ j)$ so that it is enough to consider a transposition $(i \ i+1)$. The element of N which corresponds to $(i \ i + 1)$ is given by:

$$
\sigma_{(i \ i+1)} = \begin{array}{c} \\ i \\ i+1 \end{array}
\begin{array}{cc} i & i+1 \\ & \end{array}
\begin{bmatrix}
1 & & & \vdots & & & \\
& \ddots & & \vdots & & & \\
& & 1 & \vdots & & & \\
& & & 0 & 1 & & \\
& & & 1 & 0 & & \\
& & & & & 1 & \\
& & & & & & \ddots & \\
& & & & & & & 1
\end{bmatrix}
\tag{C.21}
$$

Note that $\sigma_{(i\ i+1)}^{-1} = \sigma_{(i\ i+1)}$ and that $\sigma_{(i\ i+1)}\text{diag}[d_1,\dots,d_n]$
$\sigma_{(i\ i+1)}^{-1} = \text{diag}[d_1,\dots,d_{i-1},d_{i+1},d_i,d_{i+2},\dots,d_n]$. It follows that

$$\sigma\text{diag}[d_1,\dots,d_n]\sigma^{-1} = \text{diag}[d_{\sigma(1)},\dots,d_{\sigma(n)}] \qquad (C.22)$$

for any σ in S_n. It is, in fact, not hard to show that $N = S_n \times D_0(n,k)$ where $D_0(n,k) = D(n,k) \cap GL(n,k)$.

Now let $R = k[X_1,\dots,X_N]$ where the X_i are vectors of independent variables, $X_i = (x_{i1},\dots,x_{im})$. The symmetric group S_n acts on R by permuting the vectors X_i. In other words,

$$(\sigma \cdot f)(X_1,\dots,X_n) = f^\sigma(X_1,\dots,X_n) = f(X_{\sigma(1)},\dots,X_{\sigma(n)}) \quad (C.23)$$

where $f \in R$ and $\sigma \in S_n$. By analogy with proposition C.7, we want to find the generators of R^{S_n}. Our treatment is based on [N-2]. If ψ is a monomial in the X_i, we let $O(\psi)$ be the sum of the elements in the orbit of ψ.

Lemma C.24. *Suppose that k has characteristic 0. Then R^{S_n} is generated by elements of the form $O(\psi)$ where $\psi = x_{11}^{i_1}\cdots x_{1m}^{i_m}$ is a monomial in X_1.*

Proof: If $\psi_1,\dots,\psi_s$ are monomials in $X_1,\dots,X_s$, respectively, then the *length* of $O(\psi_1\cdots\psi_s)$ is s. Clearly any element of length 1 can be obtained from monomials in X_1. We now use induction on s. Supposing the result for all elements of length s or less, we have for any monomials $\psi_1,\dots,\psi_{s+1}$ in $X_1,\dots,X_{s+1}$, respectively,

$$O(\psi_1)O(\psi_2\cdots\psi_{s+1}) = cO(\psi_1\psi_2\cdots\psi_{s+1}) + \delta \qquad (C.25)$$

where c is a positive integer and δ consists of elements of length $\leq s^*$. The lemma follows by the induction hypothesis.

Lemma C.26. *Suppose that k has characteristic 0. The elements $O(x_{1i_1}x_{2i_2}\cdots x_{si_s})$, $s \leq n$, generate the vector subspace of R^{S_n} consisting of elements with total degree $\leq n$.*

---

* For example, if $\psi_1 = \psi_2 = \cdots = \psi_r$, $\psi_{r+j} \neq \psi_1, j = 1\dots,s+1-r$, then $c = r$.

Proof: By the previous lemma, it is enough to show that the elements $O(\psi)$, ψ a monomial in X_1 of total degree $\leq n$, can be written in terms of the $O(x_{1i_1} x_{2i_2} \cdots x_{si_s})$. We observe that $O(x_{1i_1} \cdots x_{si_s}) = c_1 O(x_{1i_1}) O(x_{2i_2} \cdots x_{si_s}) - \sum_{j=2}^{s} c_j O(x_{2i_2} \cdots x_{j-1i_{j-1}} x_{ji_1} x_{ji_j} \cdots x_{si_s})$ where the c_j are positive. Continuing we have, $O(x_{1i_1} \cdots x_{si_s}) = \delta_\ell + (-1)^{k-\ell} \Sigma_j c_j^\ell O(\psi_{j1} \psi_{j2} \cdots \psi_{j\ell})$ where δ_ℓ is a sum of products of shorter terms, the c_j^ℓ are positive, and ψ_{ji} is a monomial in X_i. Finally, for $\ell = 1$, we have

$$O(x_{1i_1} \cdots x_{si_s}) = \delta_1 + (-1)^{s-1} c^1 O(x_{1i_1} \cdots x_{1i_s})$$

where δ_1 is a sum of products of shorter terms and c^1 is positive. By induction and the fact that c^1 is positive, the result follows.

Corollary C.27. *Suppose that k has characteristic 0. Then R^{S_n} is generated by the multilinear terms $O(x_{1i_1} \cdots x_{si_s}), s \leq n$.*

Proof: Let $n_1 > n$ and let $R_1 = k[X_1, \ldots, X_N]$. There is a homomorphism $\alpha : R_1 \to R$ with $\alpha(X_i) = X_i$, $i = 1, \ldots, n$ and $\alpha(X_{n+j}) = 0, j = 1, \cdots, n_1 - n$. This determines a homomorphism $\alpha_1 : R_1^{S_{n_1}} \to R^{S_n}$ which sends $O(\psi_1 \cdots \psi_s)$ into $O(\psi_1 \cdots \psi_s)$ for multilinear monomials $\psi_1, \ldots, \psi_s$ with $s \leq n$ and sends $O(\psi_1 \cdots \psi_s)$ into 0 for monomials with $s > n$.

Example C.28. Let $n = 2$, $m = 2$ so that $X_1 = (x_{11}, x_{12})$, $X_2 = (x_{21}, x_{22})$ and $R = k[X_1, X_2]$. Then R^{S_2} is generated by the terms $O(x_{11}) = x_{11} + x_{21}, O(x_{12}) = x_{12} + x_{22}, O(x_{11}x_{21}) = 2x_{11}x_{21}$, $O(x_{12}x_{22}) = 2x_{12}x_{22}, O(x_{11}x_{22}) = x_{11}x_{22} + x_{21}x_{12} = O(x_{12}x_{21})$. Observe that $O(x_{11}x_{12}) = x_{11}x_{12} + x_{21}x_{22} = (x_{11} + x_{21})(x_{12} + x_{22}) - (x_{11}x_{22} + x_{21}x_{12}) = O(x_{11})O(x_{12}) - O(x_{11}x_{22})$ and that $O(x_{11}x_{11}) = x_{11}^2 + x_{21}^2 = (x_{11} + x_{21})^2 - 2x_{11}x_{21} = O(x_{11})O(x_{11}) - O(x_{11}x_{21})$.

Appendix D. Derivations and Separability ([S-4], [Z-3])

Let k be a field and let $R = k[x]$ where x is an indeterminate. Let K be an extension field of k and suppose that $\xi \in K$ is algebraic over k. Then the set $I_k(\xi) = \{f \in R : f(\xi) = 0\}$ is a prime ideal in R (for $(f_1 f_2)(\xi) = f_1(\xi) f_2(\xi) = 0$ implies $f_1(\xi)$ or $f_2(\xi)$ is zero as K is a field). Hence, $I_k(\xi) = (p)$ where p is irreducible. Moreover, apart from an arbitrary unit factor $a \in k^*$, p is unique and so there is exactly one monic irreducible polynomial p_ξ such that $I_k(\xi) = (p_\xi)$. We call p_ξ the *minimal polynomial of ξ over k*.

Definition D.1. An irreducible polynomial f in $k[X]$ is *separable* if $f'(x) \neq 0$ where f' is the derivative of f. An element ξ of K which is algebraic over k is *separable* (or *separably algebraic*) if its minimal polynomial ξ is separable. If K is an algebraic extension of k, then K is a *separable extension* if every element ξ of K is separable.

We say that the field of k is *perfect* if either the characteristic of k is zero or if the characteristic of k is $p \neq 0$, then $k^p = k$ where $k^p = \{a^p : a \in k\}$. An algebraically closed field k is perfect since the equation $x^p - a$ has a root in k.

Proposition D.2. *If k is perfect, then every irreducible polynomial in $k[x]$ is separable.*

Proof: If k has characteristic zero and $f(x) = a_0 x^n + a_1 x^{n-1} + \cdots + a_n$, then $f'(x) = n a_0 x^{n-1} \cdots + a_{n-1} = 0$ if and only if $(n-i)a_i = 0$, $i = 0, 1, \ldots, n-1$ i.e. if and only if $a_i = 0$, $i = 0, 1, \ldots, n-1$. Thus, $f'(x) \neq 0$ if $n > 0$ and so, every irreducible f is separable.

On the other hand, if k has characteristic $p \neq 0$ and $f(x) = a_0 x^n + \cdots + a_n$, then $f'(x) = 0$ if and only if n is divisible by p and the coefficients a_i for which $n - i$ is not divisible by p are 0 i.e. $f(x)$ is a polynomial in x^p i.e. $f \in k[x^p]$. Since $k = k^p$, $k[x^p] = k^p[x^p]$ and so $f \in k^p[x^p]$ which implies $f = (f_1)^p$ where $f_1 \in k[x]$. In other words, f is not irreducible.

185

Definition D.3.   Let K be an extension field of k. Then K is a vector space over k and we say that K is a *finite extension of k* if $\dim_k K = [K : k]$ is finite i.e. K is a finite dimensional k-vector space.

We observe that if K is a finite extension of k, then K is an algebraic extension of k for if $\xi \in K$ and $[K : k] = n < \infty$, then $1, \xi, \ldots, \xi^n$ are linearly dependent over k. We also observe that if ξ is algebraic over k, then $K = k(\xi)$ is a finite extension of k since in this case $k(\xi) = k[\xi]$ ([Z-3]) and $1, \xi, \ldots, \xi^{n-1}$ are a basis of $k[\xi]$ over k where n is the degree of the minimal polynomial of ξ.

Definition D.4.   Let K be an algebraic extension of k. An element ξ of K is a *primitive* element of K/k if $K = k(\xi)$.

Theorem D.5 (Theorem of the Primitive Element). *If k is an infinite field* and K is a finite separable extension, then K/k has a primitive element.*

Proof: Let $K = k(\xi_1, \ldots, \xi_n)$ and let $x, x_1, \ldots, x_n$ be indeterminates. Set $\tilde{k} = k(x_1, \ldots, x_n)$ and $\widetilde{K} = K(x_1, \ldots, x_n)$. Then $\widetilde{K} = \tilde{k}(\xi_1, \ldots, \xi_n)$ and $\widetilde{K}$ is a finite separable extension of $\tilde{k}$. Let

$$\tilde{\xi} = \xi_1 x_1 + \cdots + \xi_n x_n \tag{D.6}$$

and let $P(x)$ be the minimal polynomial of $\tilde{\xi}$ in $\tilde{k}[x]$. Then

$$g(x_1, \ldots, x_n)P(x) = f(x, x_1, \ldots, x_n) \in k[x, x_1, \ldots, x_n] \tag{D.7}$$

where g is a common denominator of the coefficients of P. Let us set $G(x_1, \ldots, x_n) = f(\tilde{\xi}, x_1, \ldots, x_n)$ so that G is in $K[x_1, \ldots, x_n]$. Since $P(\tilde{\xi}) = 0$, $G = 0$ and so $\partial G / \partial x_i = 0$ $i = 1, \ldots, n$. It follows that

$$\xi_i \frac{\partial f}{\partial x}(\tilde{\xi}, x_1, \ldots, x_n) + \frac{\partial f}{\partial x_i}(\tilde{\xi}, x_1, \ldots, x_n) = 0 \tag{D.8}$$

for $i = 1, \ldots, n$. Note that (D.8) is an identity in $x_1, \ldots, x_n$. Since $\tilde{\xi}$ is separable, $P'(\tilde{\xi}) \neq 0$ and $\frac{\partial f}{\partial x}(\tilde{\xi}, x_1, \ldots, x_n)$ is a non-zero polynomial

---

* If k is finite, then the theorem is still true since in that case a finite separable extension K is also a finite field and K^* is a cyclic group ([Z-3]).

in $K[x_1, \ldots, x_n]$. Since k is infinite, there are $c_1, \ldots, c_n$ in k such that

$$\frac{\partial f}{\partial x}(\xi, c_1, \ldots, c_n) \neq 0$$

$$\xi_i \frac{\partial f}{\partial x}(\xi, c_1, \ldots, c_n) + \frac{\partial f}{\partial x_i}(\xi, c_1, \ldots, c_n) = 0 \tag{D.9}$$

where $\xi = \xi_1 c_1 + \cdots + \xi_n c_n$, $i = 1, \ldots, n$. Hence, ξ is a primitive element of K.

Definition D.10. Let R be a ring and M an R-module. A mapping $D : R \to M$ is a *derivation of* R with *values in* M if (i) $D(r_1 + r_2) = D(r_1) + D(r_2)$ and (ii) $D(r_1 r_2) = r_1 D(r_2) + r_2 D(r_1)$ for all r_1, r_2 in R. The R-module of all derivations of R with values in M is denoted $\mathrm{Der}(R, M)$ and the R-module of derivations which vanish on a subring R' of R is denoted $\mathrm{Der}_{R'}(R, M)$.

We note that if R is an integral domain with quotient field $K = K(R)$ and L is a field containing R, then every $D \in \mathrm{Der}(R, L)$ extends in a unique way to a $D' \in \mathrm{Der}(K, L)$ by setting $D'(r/s) = (sD(r) - rD(s))/s^2$ for $r/s \in K$, $r, s \in R$, $s \neq 0$.

Proposition D.11. *Let $K = k(\xi)$ be a simple separable algebraic extension of K. Then every $D \in \mathrm{Der}(k, L)$ (where L is an extension field of K) may be extended in a unique way to a $D' \in \mathrm{Der}(K, L)$.*

Proof: Let p be the minimal polynomial of ξ in $k[x]$. If D' is an extension of D, then, for every polynomial f in $k[x]$, $D'(f(\xi)) = f^D(\xi) + (D'\xi)(\frac{\partial f}{\partial x})(\xi)$ where $f^D(x) = \Sigma D(a_i)x^i$ when $f(x) = \Sigma a_i x^i$. Since $D'(0) = 0$, there is at most one derivation D' of $k[\xi]$ (and hence of $k(\xi)$) extending D. Since ξ is separable, $p'(\xi) \neq 0$ and if we set $D'\xi = -p^D(\xi)/p'(\xi)$ and $D' = D$ on k, then D' determines an extension of D to K.

Corollary D.12. *If K/k is a finite separable extension, then every derivation of k extends in a unique way to K.*

Corollary D.13. *If $K = k(\xi_1, \ldots, \xi_n)$ is a separable (algebraic) extension, then every derivation of k extends in a unique way to K.*

Proof: It is enough to show that $[K : k]$ is finite. This will follow by induction on n and the lemma D.14.

Lemma D.14. *Let $k \subset K \subset L$ be a tower of fields with $[K : k] = n$ and $[L : K] = m$. Then $[L : k] = mn$.*

Proof: Let $u_1, \ldots, u_n, v_1, \ldots, v_m$ be bases of K/k and L/K respectively. Then $u_i v_j$, $i = 1, \ldots, n$, $j = 1, \ldots, m$ clearly form a basis of L/k.

Now let $K = k(x_1, \ldots, x_n)$ be a finitely generated extension of k and suppose that tr. deg $K/k = \nu$.

Definition D.15.   If K is a separable extension of $k(z_1, \ldots, z_\nu)$, then $\{z_1, \ldots, z_\nu\}$ is called a *separating transcendence basis*. We say that K/k is *separably generated* (or *separable*) if there is a separating transcendence basis.

We now have:

Theorem D.16.   If K/k is separably generated, then $\nu =$ tr. deg $K/k = \dim_K \mathrm{Der}_k(K, K)$.

Proof: Let $\{z_1, \ldots, z_\nu\}$ be a separating transcendence basis and let $K' = k(z_1, \ldots, z_\nu)$ so that K is a finite separable (algebraic) extension of K'. Since every $D \in \mathrm{Der}_k(K', K)$ is determined by its values $D(z_i)$, we have $\dim_{K'} \mathrm{Der}_k(K', K) = \nu$.* In view of corollary D.13, every such D extends in a unique way to K and so, $\dim_K \mathrm{Der}_k(K, K) \geq \nu$. Now let $D \in \mathrm{Der}_k(K, K)$ and note that $K = k(z_1, \ldots, z_\nu, \xi) = K'(\xi)$ by the theorem of the primitive element. Let $D(z_i) = a_i$ and let D_i be the derivations of K with $D_i(z_j) = \delta_{ij}$. Then $D - \Sigma a_i D_i$ vanishes on K' and so, is an extension of the 0 derivation on K'. It follows from corollary D.12 that $D - \Sigma a_i D_i = 0$ and so, $\dim_K \mathrm{Der}(K, K) \leq \nu$.

We conclude with the following:

Theorem D.17. *If $K = k(x_1, \ldots, x_n)$ and k is perfect, then K/k is*

---

* A basis is given by the "partial derivatives" D_i where $D_i(z_j) = \delta_{ij}$.

separably generated.

Corollary D.18. *If k is algebraically closed and $K = k(V)$ is the function field of an affine variety V, then $k(V)/k$ is separably generated.*

A proof of the theorem can be found in [Z-3] and shall not be given here.

Problems

1. Show that if A is an $n \times n$ matrix, then $(zI - A)^{-1} = \sum_{j=1}^{n}$ $\phi_j(z)A^{n-j}/\det[zI - A]$ where $\phi_j(z)$ is a polynomial of degree $n - j$. Compute $\phi_j(z)$.

2. Prove that if $H = (h_{i+j-1})_{i,j=1}^{\infty}$ has rank n, then there is a unique recurrence relation of the form $a_0 h_j + a_1 h_{j+1} + \cdots + a_{n-1}h_{j+n-1} + h_{j+n} = 0$ with $a_0, a_1, \ldots, a_{n-1}$ in $\mathbb{C}$, $j = 1, \ldots$. [Hint: consider the columns of H one by one.]

3. Let

$$
A = \begin{bmatrix} 0 & 1 & 0 & \cdots & 0 \\ 0 & 0 & 1 & \cdots & 0 \\ \vdots & \vdots & & & \vdots \\ \vdots & \vdots & & & 1 \\ -a_0 & -a_1 & & & -a_{n-1} \end{bmatrix}, \quad b = \begin{bmatrix} 0 \\ 0 \\ \vdots \\ 1 \end{bmatrix}
$$

and $x = (zI - A)^{-1}b$. Show that $zx_j = x_{j+1}, j = 1, \ldots, n-1$ and $x_1 = 1/\det[zI - A]$. Compute $\det[zI - A]$.

4. Let R be an integral domain. Show that $R[[x]]$ is an integral domain.

5. Show that if $H = (h_{i+j-1})_{i,j=1}^{\infty}$ has rank n, then H_{n+j} has rank n for $j = 0, 1, \ldots$, and conversely.

6. Let $\psi : \text{Hank}(n, k) \to \mathbb{A}_k^{2n}$ be given by $\psi(H) = (h_1, \ldots, h_{2n})$, and let $\mathcal{H}$ be the subset of $\mathbb{A}_k^{2n}$ given by $\mathcal{H} = \{(h_1, \ldots, h_{2n}) : H_n = (h_{i+j-1})_{i,j=1}^{n}$ is non-singular $\}$. Show that ψ is a bijective map of $\text{Hank}(n, k)$ onto $\mathcal{H}$.

7. Let A, b, c be given by equation (2.10) and let $h_j = h_j(b_0, \ldots, b_{n-1}, a_0, \ldots, a_{n-1})$ be given by (2.9). Show that $\det J(h_1,$

$\ldots, h_{2n}; b_0, \ldots, b_{n-1}, a_0, \ldots, a_{n-1})$ is not zero if and only if det $[c' \ A'c' \ \cdots \ A'^{n-1}c'] \neq 0$ and that $\det J(h_1, \ldots, h_{2n}; b_0, \ldots, b_{n-1}, a_0, \ldots, a_{n-1})$ is not zero if and only if the determinant of the matrix

$$\left.\begin{bmatrix} b_0 & b_1 & \cdots & b_{n-1} & & & \\ 0 & b_0 & \cdots & \cdots & b_{n-1} & & \\ \vdots & \vdots & & \vdots & \vdots & & \\ a_0 & a_1 & \cdots & a_{n-1} & 1 & & \\ 0 & a_0 & \cdots & \cdots & a_{n-1} & 1 & \\ \cdots & \cdots & \cdots & \cdots & \cdots & \cdots & \end{bmatrix}\begin{array}{l} \left.\vphantom{\begin{matrix}a\\a\\a\end{matrix}}\right\} n \text{ rows} \\[2em] \left.\vphantom{\begin{matrix}a\\a\end{matrix}}\right\} n-1 \text{ rows} \end{array}\right.$$

is not zero.

8.   Let R be a unique factorization domain and let $r, s \in R$. An element d of R is a *greatest common divisor* of r and s if (i) d is a divisor of r and s, and (ii) if d_1 is a divisor of r and s, then d_1 divides d. Show that any two elements r, s of R have a greatest common divisor which is unique to within a unit. Generalize to any finite number of elements of R. An element m of R is a *least common multiple* of r and s if (i) m is a multiple of r and s, and (ii) if m_1 is a multiple of r and s, then m_1 is a multiple of m. Show that any two elements r, s of R have a least common multiple which is unique to within a unit. Generalize to any finite number of elements of R. Show that if d is a greatest common divisor and m a least common multiple of r, s, then $m \cdot d = rs$.

9.   Let $R = Z$ be the integers and let $f(x) = 3x^4 + x^3 + 7$, $g(x) = 2x^2 + x + 5$. What are $d(x)$, $r(x)$? Develop several "algorithms" for computing $d(x), r(x)$ for general $f(x), g(x)$ in $Z[x]$. Evaluate the "algorithms" in terms of computational efficiency.

10.   Let $p(z) = b_0 + b_1 + b_1 z + \cdots + b_{n-1} z^{n-1}$, $q(z) = a_0 + a_1 z + \cdots + a_{n-1} z^{n-1} + z^n$. Show that the resultant of p and q, $\text{Res}(p, q)$, is given by

$$\det \left.\begin{bmatrix} b_0 & b_1 & \cdots & b_{n-1} & & & \\ 0 & b_0 & \cdots & \cdots & b_{n-1} & & \\ \vdots & \vdots & & \vdots & \vdots & & \\ a_0 & a_1 & \cdots & a_{n-1} & 1 & & \\ 0 & a_0 & \cdots & \cdots & a_{n-1} & 1 & \\ \cdots & \cdots & \cdots & \cdots & \cdots & \cdots & \end{bmatrix}\begin{array}{l} \left.\vphantom{\begin{matrix}a\\a\\a\end{matrix}}\right\} n \text{ rows} \\[2em] \left.\vphantom{\begin{matrix}a\\a\end{matrix}}\right\} n-1 \text{ rows} \end{array}\right.$$

11. Let $f(x) = a_m x^m + \cdots + a_0$, $g(x) = b_n x^n + \cdots + b_0$. Show that $\mathrm{Res}(f, g)$ is a homogeneous polynomial in the coefficients a_i, b_j of degree $m + n$.

12. Show that a field k is algebraically closed (i) if and only if every prime in $k[x]$ has degree 1; or, (ii) if and only if every $f(x)$ in $k[x]$ factors completely into linear factors; or, (iii) if and only if every $f(x)$ in $k[x]$ of positive degree has a root in k.

13. Let R, S be rings and let $\psi : R \to S$ be a homomorphism. Show that Ker $\psi = \mathfrak{a}$ is an ideal (Ker ψ is the kernel of ψ). If $R = Z$, $S = Z$ and $\psi : R \to S$ is given by $\psi(r) = 0$ or 1 according as r is even or odd, show that ψ is a homomorphism. What is Ker ψ?

14. Let $\mathfrak{a}, \mathfrak{a}_1$ be ideals in $k[x_1, \ldots, x_N]$ and let W, W_1 be subsets of $\mathbb{A}_k^N$. Show that if $\mathfrak{a} \subset \mathfrak{a}_1$, then $V(\mathfrak{a}) \supset V(\mathfrak{a}_1)$ and if $W \subset W_1$, then $I(W) \supset I(W_1)$.

15. Determine the Zariski closed sets in $\mathbb{A}_k^1$ and in $\mathbb{A}_k^2$.

16. Show that every ideal in $k[x]$ is principal.

17. Give a counterexample to lemma 5.8 when k is not algebraically closed.

18. Let R, S be rings and let $\psi : R \to S$ be a homomorphism. Show that if $\mathfrak{p}$ is a prime ideal in S, then $\psi^{-1}(\mathfrak{p})$ is a prime ideal in R. Suppose that $S = R/\mathfrak{a}$ where $\mathfrak{a}$ is an ideal in R and ψ is the natural homomorphism. Show that the prime ideals of S are the ideals of the form $\mathfrak{p}/\mathfrak{a}$ where $\mathfrak{p}$ is a prime ideal in R and that the maximal ideals in S are the ideals of the form $\mathfrak{m}/\mathfrak{a}$ where $\mathfrak{m}$ is a maximal ideal in R.

19. Let R be an affine k-algebra and let $\mathrm{Spm}(R) = \{\mathfrak{m} : \mathfrak{m}$ is a maximal ideal in $R\}$. If $\mathfrak{a}$ is an ideal in R, let $V(\mathfrak{a}) = \{\mathfrak{m} \in \mathrm{Spm}(R) : \mathfrak{a} \subset \mathfrak{m}\}$. Show that the $V(\mathfrak{a})$ define a topology (the Zariski topology) on $\mathrm{Spm}(R)$.

20. Let $\mathfrak{a}$ be an ideal in $k[X_1, \ldots, X_N]$ and let $V(\mathfrak{a}) = \{M_\xi : M_\xi \supset \mathfrak{a}\}$. Show that $\sqrt{\mathfrak{a}} = \cap M_\xi$ where $M_\xi \in V(\mathfrak{a})$. If R is an affine

k-algebra and f is an element of R such that $\alpha(f) = 0$ for all $\alpha \in \operatorname{Hom}(R, k)$, then $f = 0$.

21.   Let $F : \mathbb{A}_k^N \to \mathbb{A}_k^M$ be a polynomial mapping. Show that if $V \subset \mathbb{A}_k^N$ is irreducible, then $F(V)$ is irreducible. What does this mean for morphisms?

22.   Let $\tilde{L}$ be given by equations (8.3), (8.4), (8.5). Show that $\tilde{L}(\operatorname{Rat}(n, k)) \subseteq \operatorname{Hank}(n, k)$. Let $\tilde{F}$ be given by equations (8.6)–(8.10). Show that $\tilde{F}(\operatorname{Hank}(n, k)) \subseteq \operatorname{Rat}(n, k)$.

23.   Let V be an affine variety with $\mathfrak{p} = I(V)$ and let $\xi \in V$. Show that $k[X_1, \ldots, X_N]_{M_\xi}/\mathfrak{p}k[X_1, \ldots, X_N]_{M_\xi}$ is isomorphic to $k[V]_{\mathfrak{m}_\xi} = (k[X_1, \ldots, X_N]/\mathfrak{p})_{M_\xi/\mathfrak{p}}$. (In other words, both define $\mathcal{O}_{\xi, V}$.)

24.   Let V be an affine variety and let $\xi \in V$. Consider the set of pairs (f, U) where f is regular on U (in the sense of definition 9.8) and U is an open subset of V with $\xi \in U$. Say that $(f_1, U_1)E_\xi(f_2, U_2)$ if $f_1 = f_2$ on $U_1 \cap U_2$. Show that E_ξ is an equivalence relation and that the set of equivalence classes, with the natural definitions of addition and multiplication, is a local ring isomorphic to $\mathcal{O}_{\xi, V}$.

25.   Let $V \subset \mathbb{A}_k^N$, $W \subset \mathbb{A}_k^M$ be quasi-affine varieties and let $\phi : V \to W$ be a continuous map. Then ϕ is a morphism if and only if $\phi^*(\mathcal{O}_{\phi(\xi), W}) \subset \mathcal{O}_{\xi, V}$ for all $\xi \in V$.

26.   Prove proposition 10.8.

27.   Let $\operatorname{Char}(n, k) = \{\xi \in \mathbb{A}_k^{2n} : \lambda(\xi) \neq 0\} = \psi_\chi^{-1}(\operatorname{Rat}(n, k))$. Then show that if $x = (A, b, c)$ is an element of $S_{1,1}^n$, then $\mathfrak{R}_\chi(x) = \mathfrak{R}_\chi(A, b, c)$ is an element of $\operatorname{Char}(n, k)$ and conversely, if ξ is an element of $\operatorname{Char}(n, k)$, then there is an (A, b, c) in $S_{1,1}^n$ with $\mathfrak{R}_\chi(A, b, c) = \xi$.

28.   Let $n = 2$ and let X, Y, Z be the coordinate functions on $\mathbb{A}_k^6$ (i.e. $X = (X_{ij})_{i,j=1}^2$, etc.). Set $\lambda(X, Y, Z) = \lambda(\mathfrak{R}_\chi(X, Y, Z))$, $\theta(X, Y, Z) = \theta(\mathfrak{R}_h(X, Y, Z))$, and $\rho(X, Y, Z) = \rho(\mathfrak{R}_f(X, Y, Z))$. Show by direct computation that $\lambda = \rho = -\theta$.

29.   Show that the controllability and observability maps are not

surjective. What about the Hankel matrix map?

30.   Let $k = \mathbb{C}$, the field of complex numbers. Suppose that (A, b, c) is controllable and consider the differential equation $\dot{x} = Ax + bu$. Show that given any x_0 in $\mathbb{C}^n$, there is a $u(t)$ such that the solution $x(t; x_0, u(\cdot))$ of the differential equation with $x(0; x_0, u(\cdot)) = x_0$ passes through 0 for some $t > 0$.

31.   Let $k = \mathbb{C}$, the field of complex numbers. Suppose that (A, b, c) is observable and consider the system $\dot{x} = Ax + bu$, $y = cx$. Show that, given $u(\cdot)$ and $y(\cdot)$ on an interval $(0, t]$ with $t > 0$, then the initial condition x_0 such that $y(\cdot) = cx(\cdot; x_0, u(\cdot))$ can be determined.

32.   Let $G = GL(n, k)$. Show that G is a group and that the mappings $\alpha : G \times G \to G$ and $\beta : G \to G$ given by $\alpha(g_1, g_2) = g_1 g_2$, $\beta(g) = g^{-1}$ are regular (where $G \times G$ is viewed as a subset of $\mathbb{A}_k^{2n^2}$).

33.   Let $V \subset \mathbb{A}_k^N$ and $W \subset \mathbb{A}_k^M$ be affine varieties. Show that $\{\xi\} \times W$ is isomorphic to W and that $V \times \{\eta\}$ is isomorphic to V. What if V and W are algebraic sets?

34.   Let $n = 2$ and $G = GL(2, k)$. Consider the action of G on $\mathbb{A}_k^8$ given by $g \cdot (A, b, c) = (gAg^{-1}, gb, cg^{-1})$. Let $x_0 = (A_0, b_0, c_0)$ with

$$A_0 = \begin{bmatrix} 0 & 1 \\ 0 & 0 \end{bmatrix} \quad , \quad b_0 = \begin{bmatrix} 0 \\ 1 \end{bmatrix} \quad , \quad c_0 = \begin{bmatrix} 1 & 0 \end{bmatrix}$$

What is the stabilizer subgroup $S(x_0)$? Let $x_1 = (A_1, b_1, c_1)$ with

$$A_1 = \begin{bmatrix} 1 & 0 \\ 0 & 1 \end{bmatrix} \quad , \quad b_1 = \begin{bmatrix} 0 \\ 1 \end{bmatrix} \quad , \quad c_1 = \begin{bmatrix} 1 & 0 \end{bmatrix}$$

What is the stabilizer subgroup $S(x_1)$? Explain the results.

35.   Let $R = k[S_{1,1}^n]$ be the ring of regular functions on $S_{1,1}^n$. Viewing the elements of R as functions from $S_{1,1}^n$ into $\mathbb{A}_k^1$, show that the action of G defines a map (or action) from $G \times R \to R$ and that the set of invariant elements of R forms a subring R^G of R. In fact, show also that the action of G on $\mathbb{A}_k^{n^2+2n}$ defines an action from $G \times k[\mathbb{A}_k^{n^2+2n}] \to k[\mathbb{A}_k^{n^2+2n}]$ which preserves degrees.

36.   Let E be an equivalence relation on V and let $\mathcal{F}_E = \{\mathfrak{a} : \mathfrak{a}$ is an E-invariant ideal in $k[V]\}$. Consider the set $\{V(\mathfrak{a}) : \mathfrak{a} \in \mathcal{F}_E\}$. Does this set define a topology on V? If so, what can you say about f in R^E?

37.   Let E be an equivalence relation on V and let $R = k[V]$, $R^E = k[V]^E$. Then there is a natural map $\psi : \mathrm{Spm}(R) \to \mathrm{Id}(R^E)$ given by $\psi(\mathfrak{m}) = \mathfrak{m} \cap R^E$ where $\mathrm{Id}(R^E)$ is the set of ideals in R^E. Show that ψ is an invariant and that $\mathrm{Spm}(R^E)$ is contained in the range of ψ.

38.   Let E be an equivalence relation on a variety V and let $R = k[V]$, $R^E = k[V]^E$. Suppose that $f \in R^E$ i.e. f is an invariant. Show that $(R_f)^E = (R^E)_f$. If a variety W and a morphism ψ determine a geometric quotient of V modulo E, show that, for each open set $U_0 \subset W$, $\psi^* : \mathcal{O}_W(U_0) \to \mathcal{O}_V(\psi^{-1}(U_0))$ is a surjective k-isomorphism between $\mathcal{O}_W(U_0)$ and the ring of invariants on $\psi^{-1}(U_0)$, $\mathcal{O}_V(\psi^{-1}(U_0))^E$. [Hint: treat $U_0 = W_f$ with $f \in k[W]$ first.]

39.   Let $R = k[X, Y, Z]$ $(X = (X_{ij}), Y = (Y_j), Z = (Z_i))$ and let $G = GL(n, k)$ act on R via $g \cdot (X, Y, Z) = (gXg^{-1}, gY, Zg^{-1})$. Show that, for fixed g, the map $\psi_g : R \to R$ given by $\psi_g(p(X, Y, Z)) = p(gXg^{-1}, gY, Zg^{-1})$ is a k-automorphism of R.

40.   Develop analogs of proposition 15.4 and corollary 15.5 for the cases of $\mathrm{Rat}(n, k)$ and $\mathrm{Hank}(n, k)$.

41.   (Linear and Algebraic dependence, [Z-3]). Let X be a set and let $2^X = \{A : A \subset X \text{ (i.e. } A \text{ is a subset of } X)\}$. A mapping $s : 2^X \to 2^X$ is a *span* if: (i) $A \subset B$ implies $s(A) \subset s(B)$; (ii) if $x \in X$, $x \in s(A)$, then there is a finite $A' \subset A$ with $x \in s(A')$; (iii) $A \subset s(A)$; (iv) $s(A) = s(s(A))$; (v) if $x \in s(A \cup \{a\}), x \notin s(A)$, then $a \in s(A \cup \{x\})$. Say that A *generates* X if $s(A) = X$; that A is *free* (*independent*) if $a \notin s(A - \{a\})$ for all $a \in A$; and that A is a *basis* if A is free and generates X. Show that A is a basis if and only if A is a minimal set of generators or A is a maximal free subset. Show that (a) every free set A extends to a basis; (b) every set of generators contains a basis; and, (c) if X has a finite basis, then all bases are finite and have the same number of elements. Show that if X is a k-vector space and $s(A) = \mathrm{span}_k\{A\}$, then s is a span. Show that if K/k is field extension and, for $A \subset K$, $s(A)$ is the algebraic closure of $k(A)$ in K, then s is a span. Interpret the results

in terms of transcendence for this case.

42.   Give an example which shows that $V_1 < V_2$ does not imply $\dim_k V_1 < \dim_k V_2$.

43.   Let R be a ring. Show that R, $R[x]$, $R[[x]]$ are R-modules.

44.   Let R be a ring and let $R^n = R \oplus \cdots \oplus R$ (n summands). Prove that M is a finitely generated R-module if and only if M is isomorphic to a quotient of R^n.

45.   Let M be a finitely generated R-module, $\mathfrak{a}$ be an ideal in R, and $\phi : M \to M$ an R-module morphism such that $\phi(M) \subset \mathfrak{a}M$. Show that $\phi^n + a_1\phi^{n-1} + \cdots + a_n = 0$ with $a_i \in \mathfrak{a}$ (cf. proof of proposition 16.19). Show that if $\mathfrak{b}$ is an ideal in R with $\mathfrak{b}M = M$, then there is an $r \in R$ such that $r - 1 \in \mathfrak{b}$ and $rM = 0$.

46.   Let P be a multiplicatively closed set in the integral domain R. If $\mathfrak{a}$ is an ideal in R, then $R_P\mathfrak{a}$ is a proper ideal in R_P if and only if $\mathfrak{a} \cap P = \phi$ (i.e. $\mathfrak{a}$ does not meet P).

47.   (Field Polynomial, Norm and Trace [L-1], [Z-3]). Let $k \subset K$ be fields and suppose that K is a finite extension of k i.e. K is a finite dimensional vector space over k with $[K : k] = \dim_k K = n$. In other words, K is an algebraic extension of k of finite degree. If $\xi \in K$, then the mapping $\mathbf{A}_\xi : K \to K$ given by $\mathbf{A}_\xi(w) = \xi \cdot w$, $w \in K$ is a k-linear mapping. Let $w_1, \ldots, w_n$ be a basis of K/k and let A_ξ be the matrix of $\mathbf{A}_\xi$ with respect to this basis. Set $F_\xi(X) = \det(x \cdot I - A_\xi)$ and call $F_\xi(X)$ the *field polynomial of ξ over k*. Show that $F_\xi(X)$ is independent of the choice of basis and that $F_\xi(\xi) = 0$. Let $\text{Tr}_{K/k}(\xi) = $ trace of A_ξ and $N_{K/k}(\xi) = $ determinant of A_ξ and call $\text{Tr}_{K/k}(\xi)$ the *trace of ξ over k* and $N_{K/k}(\xi)$ the *norm of ξ over k*. Show that (i) $\text{Tr}_{K/k}(\xi + \eta) = \text{Tr}_{K/k}(\xi) + \text{Tr}_{K/k}(\eta)$; (ii) $N_{K/k}(\xi\eta) = N_{K/k}(\xi)N_{K/k}(\eta)$, and (iii) $N_{K/k}(\xi) \neq 0$ if $\xi \neq 0$. Suppose that $k \subset K \subset L$ is a tower of finite extensions with $[K : k] = n$, $[L : K] = m$. Show that $[L : k] = nm$ (cf. sec 4). If $\xi \in K$, then ξ has a field polynomial $F_{\xi,K/k}(X)$ as an element of K and ξ has a field polynomial $F_{\xi,L/k}(X)$ as an element of L. Show that

$$F_{\xi,L/k}(X) = (F_{\xi,K/k}(X))^m$$

and hence, that $\text{Tr}_{L/k}(\xi) = m\text{Tr}_{K/k}(\xi)$ and $N_{L/k}(\xi) = (N_{K/k}(\xi))^m$ $= (N_{K/k}(\xi))^{[L:k]}$. Apply this to the situation $k \subset k(\xi) \subset L$ to show that the field polynomial of ξ is a multiple of the minimal polynomial of ξ. Interpret this for norms.

48. Let R be a ring. Let $\mathfrak{p}_1, \ldots, \mathfrak{p}_n$ be prime ideals in R and let $\mathfrak{a}$ be an ideal contained in $\cup_1^n \mathfrak{p}_i$. Show that $\mathfrak{a} \subset \mathfrak{p}_i$ for some i. [Hint: use induction on n in the form $\mathfrak{a} \not\subset \mathfrak{p}_i$, $i = 1, \ldots, n$ implies that $\mathfrak{a} \not\subset \cup_{i=1}^n \mathfrak{p}_i$.]

49. Prove lemma 18.16. [Hint: apply the Noether Normalization lemma 16.43.]

50. Let R be an integral domain with quotient field K and let R' be the integral closure of R in K. If f, g are monic polynomials in $K[X]$ such that $fg \in R'[X]$, then $f, g \in R'[X]$. [Hint: let K_1 be a field with $K \subset K_1$ such that f, g split into linear factors i.e. $f = \prod(X - \alpha_i)$, $g = \prod(X - \beta_j)$, $\alpha_i, \beta_j \in K_1$. Then α_i, β_j are integral over R' being roots of fg and so, the coefficients of f and g are integral over R'.] Can you prove a generalization when R is not an integral domain? ([A-2], [B-2]).

51. Prove the results of 19.16-19.22.

52. Show that if $f \in S = R^G$, then $(S_{1,1}^n)_f$ is an invariant open set in $S_{1,1}^n$. Show also that if f is a prime in $R = k[X, Y, Z]$ and $(S_{1,1}^n)_f$ is invariant, then $f \in S$ i.e. f is invariant (cf. [B-7]).

53. Show that the maps α_1, α_2 of (19.23) are regular and that $f \in S$ if and only if $(\alpha_1 \circ \alpha_2)^*(f) = f$.

54. Let $D \subset \mathbb{A}_k^{n^2+2n}$ be $V(X_{ij} : i \neq j, i, j = 1, \ldots n)$ so that $I(D) = (X_{ij})$, $i \neq j$. Show that $R_D = k[D]$ is isomorphic with $k[X_{11}, \ldots, X_{nn}, Y, Z]$. Is D irreducible? What is the dimension of D?

55. Let $U_1 = (X_{11}, W_1)$, $U_2 = (X_{22}, W_2)$, $U_3 = (X_{33}, W_3)$. Determine the generators of $k[U]^{S_3}$ explicity. Show that $S_{0D} \cap k[U]^{S_3} \neq k[U]^{S_3}$ (using 19.27) and that $S_D = S_{0D}$ by direct calculation. Show also that S_{0D} is generated by $O(X_{11})$, $O(X_{11}^2)$, $O(X_{11}^3)$, $O(W_1)$, $O(X_{11}W_1)$, and $O(X_{11}^2 W_1)$. Show also that $O(X_{11}W_1^2)$ is not an element of S_D.

56. Prove that $\text{Hank}(n, k)$ and $\mathcal{H}$ are isomorphic.

57.   Show that $A(N, k)$ is a group and that the maps μ : $A(N, k) \times A(N, k) \to A(N, k)$ and $\theta : A(N, k) \to A(N, k)$ given by $\mu(\alpha_1, \alpha_2) = \alpha_1 \alpha_2$, $\theta(\alpha) = \alpha^{-1}$ are regular. Show also that $A(N, k)$ acts on $\mathbb{A}_k^N$ (a fortiori on $k[X_1, \ldots, X_N]$) and that the action on $k[X_1, \ldots, X_N]$ preserves degrees.

58. Show that if V is an affine algebraic set, then so is $\alpha(V)$ for $\alpha \in A(N, k)$ and that V and $\alpha(V)$ are isomorphic.

59.   Let $V = V(\ell_1 - a_1, \ldots, \ell_m - a_m)$ be a linear variety in $\mathbb{A}_k^N$. Prove that there is an affine transformation $\alpha = (g, \xi_0)$ such that if $Z_i = (gX + \xi_0)_i$, then $Z_i = \ell_i(X) - a_i$ for $i = 1, \ldots, m$.

60. Let V be an affine variety with $\mathfrak{p} = I(V)$. Let $f_1, \ldots, f_r$ be a basis of $\mathfrak{p}$ and let $\xi \in V$. Show that rank $J(f_1, \ldots, f_r; X_1, \ldots X_N)$ (ξ) is independent of the choice of basis of $\mathfrak{p}$.

61. Let $\psi : V \to W$ be a morphism of affine varieties. Show that the tangent map is a homomorphism. What form does the tangent map take in terms of bases of $I(V)$ and $I(W)$?

62.   Determine the set $\text{Sing}(V)$ for $V = V(X_1^2 - X_2^3)$ in $\mathbb{A}_{\mathbb{C}}^2$. Determine $T_{V, \xi}$ for all points ξ of V.

63. Prove proposition 20.40 in detail.

64. Let Γ be a closed subgroup of $GL(n, k)$. Show that there is a unique irreducible component Γ_e of Γ containing the identity I and that Γ_e is a closed normal subgroup of Γ. Use this to show that the feedback group is irreducible. Suppose that Γ is a closed irreducible subgroup of $GL(n, k)$. Show that Γ is nonsingular and hence, that the feedback group is nonsingular.

65.   Prove that range $[b \ Ab \ \cdots \ A^{n-1}b] = $ range $[b \ (A - b\alpha^{-1}K)b \ \cdots \ (A - b\alpha^{-1}K)^{n-1}b]$.

66. Show that N_f and G_f are closed linear subvarieties of Γ_f

and that $\dim N_f = n$, $\dim G_f = n^2 + 1$. If $\gamma = [I, K, 1] \in N_f$, then show that $\gamma_0 = \gamma - [I, 0, 1]$ is nilpotent with $\gamma_0^{n+1} = 0$ (as matrices). (Such a γ is called *unipotent* ([H-8], [S-4])).

67.   Show that $\Delta(\Gamma_f, x) = \Delta(N_f, x) = \Delta(\Gamma_f, \gamma_g x) = \Delta(N_f, \gamma_g x)$ for $\gamma_g \in G_f$.

68.   Show that if k has characteristic 0, then Δ_x is surjective if and only if T_x is surjective. What is the situation if k has characteristic $p \neq 0$?

69.   Prove that if $X \in M(n, 1; k)$ and $Y \in M(1, n; k)$, then $\det[I + X \cdot Y] = 1 + Y \cdot X$ using the Binet-Cauchy formula ([W-1]).

70.   Prove that if $X \in M(n, 1; k)$ and $Y \in M(1, n; k)$, then $\det[I + X \cdot Y] = 1 + Y \cdot X$ by showing that $X \cdot Y$ is similar to a diagonal matrix with entries $[\text{Tr}(X \cdot Y), 0, \cdots, 0]$ and noting that $\text{Tr}(X \cdot Y) = \text{Tr}(Y \cdot X) = Y \cdot X$ ([W-1]).

71.   Prove corollary 21.22.

72.   Prove that $T_x(K) = T_x(0) + t_x(K)N_x(K)$ where $N_x(K)$ is given by 21.28.

73.   Suppose that k has characteristic $p \neq 0$ with $p < n$. Can T_x be surjective?

74.   Show that if $k = \mathbb{C}$ is the complex numbers, then the map $\Delta_x(K) = \det[zI - A + bK]$ is surjective if and only if the map $\Lambda_x(K) = (\lambda_1, \ldots, \lambda_n)$ is surjective where $\lambda_i = \lambda_i(x, K)$ are the roots of $\Delta_x(K)(z) = 0$. What is the situation when $k = \mathbb{R}$ is the real numbers?

75.   Prove corollary 22.11 in detail ([D-1], [S-4]).

References

[A-1] Athans, M. and Falb, P.L., *Optimal Control: An Introduction to the Theory and its Applications*, McGraw-Hill, New York, 1966.

[A-2] Atiyah, M.F. and MacDonald, I.G., *Introduction to Commutative Algebra*, Addison-Wesley, Reading, MA, 1969.

[B-1] Birkhoff, G. and Maclane, S., *A Survey of Modern Algebra*, Revised Edition, Macmillan, New York, 1953.

[B-2] Borel, A., *Linear Algebraic Groups*, Benjamin, New York, 1969.

[B-3] Bourbaki, N. *Algèbre Commutative*, I-VII, Hermann, Paris, 1961-65.

[B-4] Brockett, R. *Finite-Dimensional Linear Systems*, Wiley-Interscience, New York, 1970.

[B-5] Byrnes, C. "The moduli space for a linear dynamical system" in *Geometric Control Theory* (C. Martin, R. Hermann, eds.), Math. Sci. Press, Brookline, MA, 1977.

[B-6] Byrnes, C. and Falb, P.L., "Applications of algebraic geometry in system theory", Am. J. Math., Vol. 101, 1979.

[B-7] Byrnes, C. and Hurt, N. "On the moduli of linear dynamical systems", Adv. in Math: Studies in Analysis, Vol. 4, 1978.

[D-1] Dieudonné, J. *Cours de Géométrie Algébrique*, 1, 2, Presses Universitaires de France, 1974.

[D-2] Dieudonné, J. and Carrell, J., "Invariant theory, old and new", Advances in Math, 4, 1970.

[F-1] Falb, P.L., *Linear Systems and Invariants*, Lecture Notes, Control Group, Lund University, Sweden, 1974.

[F-2] Fogarty, J., *Invariant Theory*, Benjamin, New York, 1969.

[F-3] Fulton, W., *Algebraic Curves*, Benjamin, New York, 1969.

[G-1] Gantmacher, F.R., *Theory of Matrices* I, II, Chelsea, New York, 1959.

[H-1] Haboush, W.J., "Reductive groups are geometrically reductive", Ann. of Math, Vol. 102, 1975.

[H-2] Hartshorne, R., *Algebraic Geometry*, Springer-Verlag, Berlin-Heidelberg-NewYork, 1977.

[H-3] Hazewinkel, M., "Moduli and canonical forms for linear dynamical systems III: the algebraic-geometric case", in *Geometric Control Theory* (C. Martin, R. Hermann, eds.), Math. Sci. Press, Brookline, MA, 1977.

[H-4] Hazewinkel, M. and Kalman, R.E., "On invariants, canonical forms and moduli for linear constant finite-dimensional, dynamical systems" in *Lecture Notes Economics-Math System Theory*, Springer-Verlag, Berlin-Heidelberg-New York, 1976.

[H-5] Hermann, R. *Linear Systems Theory and Introductory Algebraic Geometry*, Math. Sci. Press, Brookline, MA, 1974.

[H-6] Hermann, R. and Martin, C., "Applications of algebraic geomtry to system theory-Part I", IEEE Trans. Aut. Cont., AC-22, 1977.

[H-7] Hochschild, G., *Basic Theory of Algebraic Groups and Lie Algebras*, Springer-Verlag, Berlin-Heidelberg-New York, 1981.

[H-8] Humphreys, J., *Linear Algebraic Groups*, Springer-Verlag, Berlin-Heidelberg-New York, 1975.

[J-1] Jacobson, N., *Lectures in Abstract Algebra* I, II, III, Van Nostrand, Princeton, NJ, 1953.

[K-1] Kalman, R.E., "Global Structure of classes of linear dynamical systems" in *Lectures*, NATO Adv. Study Inst., London, 1971.

[K-2] Kalman, R.E., Falb, P.L. and Arbib, M.A., *Topics in Mathematical System Theory*, McGraw-Hill, New York, 1969.

[K-3] Kendig, K., *Elementary Algebraic Geometry*, Springer-Verlag, Berlin-Heidelberg-New York, 1977.

[K-4] Kunz, E., *Inroduction to Commutative Algebra and Algebraic Geometry*, Birkhäuser-Boston, Boston, 1985.

[L-1] Lang, S., *Algebra*, Addison-Wesley, Reading, MA, 1971.

[M-1] Matsumura, H., *Commutative Algebra*, 2nd Ed., Benjamin, New York, 1980.

[M-2] Mumford, D.B., *Introduction to Algebraic Geometry* (Preliminary Version of First 3 Chapters), Lecture Notes, Harvard University, 1965.

[M-3] Mumford, D.B., *Geometric Invariant Theory*, Springer-Verlag, Berlin-Heidelberg-New York, 1965. [2nd Enlarged Edition, with J. Fogarty, 1982].

[M-4] Mumford, D. and Suominen, K., "Introduction to the theory of moduli" in Proc Fifth Summer School in Math., Oslo, 1970.

[N-1] Nagata, M., "Invariants of a group in an affine ring", J. Math Kyoto, 1964.

[N-2] Neeman, A., "Topics in Algebraic Geometry", Thesis, Harvard University, 1983.

[N-3] Northcott, D. G., *Affine Sets and Affine Groups*, Cambridge Univ. Press, Cambridge (Eng.), 1980.

[R-1] Rosenbrock, H., *State-space and Multivariable Theory*, Wiley-Interscience, New York, 1970.

[S-1] Samuel, P., *Méthodes d'Algèbre Abstraite en Géométrie Algébrique*, Springer-Verlag, Berlin-Heidelberg-New York, 1967.

[S-2] Shafarevich, I.R., *Basic Algebraic Geomtery*, Springer-Verlag, Berlin-Heidelberg-New York, 1977.

[S-3] Sontag, E., "Linear systems over commutative rings: a survey", Richerche di Automatica, Vol. 7, 1976.

[S-4] Springer, T.A., *Linear Algebraic Groups*, Birkhäuser-Boston, Boston, 1981.

[W-1] Wolovich, W., *Linear Multivariable Systems*, Springer-Verlag, Berlin-Heidelberg-New York, 1974.

[W-2] Wonham, W.M., "On pole assignment in multi-input controllable linear systems", IEEE Trans. Aut. Cont., AC-12, 1967.

[W-3] Wonham, W.M. and Morse, A.S., "Feedback invariants of linear multivariable systems", Automatica, Vol. 8, 1972.

[Z-1] Zariski, O., "A new proof of the Hilbert Nullstellensatz", Bull. Am. Math. Soc., Vol. 53, 1947.

[Z-2] Zariski, O., "The concept of a simple point of an abstract algebraic variety", Trans. Am. Math. Soc., Vol. 62, 1947.

[Z-3] Zariski, O and Samuel, P., *Commutative Algebra* I, II, Van Nostrand, Princeton, NJ, 1958, 1960.

Systems & Control: Foundations & Applications

Series Editor

Christopher I. Byrnes
Department of Systems Science and Mathematics
Washington University
Campus P.O. 1040
One Brookings Drive
St. Louis, MO 63130-4899
U.S.A.

Systems & Control: Foundations & Applications publishes research mono-graphs and advanced graduate texts dealing with areas of current research in all areas of systems and control theory and its applications to a wide variety of scientific disciplines.

We encourage preparation of manuscripts in such forms as LaTeX or AMS TeX for delivery in camera-ready copy which leads to rapid publication, or in electronic form for interfacing with laser printers or typesetters.

Proposals should be sent directly to the editor or to: Birkhäuser Boston Inc., 675 Massachusetts Avenue, Suite 601, Cambridge, MA 02139, U.S.A.

SC1   Estimation Techniques for Distributed Parameter Systems
       H.T. Banks and K. Kunisch

SC2   Set-Valued Analysis
       Jean-Pierre Aubin and Hélène Frankowska

SC3   Weak Convergence Methods and Singularly Perturbed Stochastic Con-trol and Filtering Problems
       Harold J. Kushner

SC4   Methods of Algebraic Geometry in Control Theory: Part I
       Scalar Linear Systems and Affine Algebraic Geometry
       Peter Falb